William Guy Peck

Elements of Mechanics

Treated by Means of the Differential and Integral Calculus

William Guy Peck

Elements of Mechanics
Treated by Means of the Differential and Integral Calculus

ISBN/EAN: 9783337811822

Printed in Europe, USA, Canada, Australia, Japan

Cover: Foto ©berggeist007 / pixelio.de

More available books at **www.hansebooks.com**

ELEMENTS

OF

MECHANICS:

TREATED BY MEANS OF

THE DIFFERENTIAL AND INTEGRAL
CALCULUS.

BY

WILLIAM G. PECK, Ph.D., LL.D.,

PROFESSOR OF MATHEMATICS, ASTRONOMY, AND MECHANICS, COLUMBIA
COLLEGE.

A. S. BARNES & COMPANY
NEW YORK AND CHICAGO

PREFACE

The following work was undertaken to supply a want felt by the author, when engaged in teaching Natural Philosophy to College classes. In selecting a text-book on the subject of MECHANICS, there was no want of material from which to choose; but to find one of the exact grade for College instruction, was a matter of much difficulty. The higher treatises were found too difficult to be read with profit, except by a few in each class, in addition to which they were too extensive to be studied, even by the few, in the limited time allotted to this branch of education. The simpler treatises were found too elementary for advanced classes, and on account of their non-mathematical character, not adapted to prepare the student for subsequent investigations in Science.

The present volume was intended to occupy the middle ground between these two classes of works, and to form a connecting link between the Elementary and the Higher Treatises. It was designed to embrace all of the important propositions of Elementary Mechanics, arranged in logical order, and each rigidly demonstrated. If these designs have been accomplished, this volume can be read with facility and advantage, not only by College classes, but by the higher classes in Academies and High Schools; it will be found to contain a sufficient amount of information for those who want either the leisure or the desire to make the mathematical sciences a specialty; and finally, it will serve as a suitable introduction to those higher treatises on Mechanical Philosophy, which all must study who would appreciate and keep pace with the wonderful discoveries that are daily being made in Science.

COLUMBIA COLLEGE, *February* 22, 1859.

PREFACE TO THE SECOND EDITION.

In accordance with the expressed wish of many teachers in institutions where the Differential and Integral Calculus are either not taught at all, or else are not obligatory studies, an Appendix has been added to the body of the work, in which all of the principles there demonstrated by means of the Calculus are deduced by the aid of Elementary Mathematics only.

It has not seemed desirable to omit the Calculus altogether, especially as by the present arrangement the work is equally adapted to the use of those who teach by the aid of the Calculus, and of those who only employ the Elementary Mathematics.

From the flattering reception of this work by the Public, it is believed that a continuation of the Course of Natural Philosophy, of which this is the opening volume, would be acceptable. To carry out this design, two other volumes are in preparation on the same general plan as the present, one of which will be devoted to the subjects of *Acoustics* and *Optics*, and the other to *Heat and the Steam-Engine*, *Electricity*, and *Magnetism*.

FEBRUARY 22, 1860.

CONTENTS.

CHAPTER I.

	PAGE
DEFINITIONS—Rest and Motion	13
Forces	14
Gravity	15
Weight—Mass	16
Momentum—Properties of Bodies	17
Definition of Mechanics—Measure of Forces	21
Representation of Forces	23

CHAPTER II.

Composition of Forces whose Directions coincide	25
Parallelogram of Forces	26
Parallelopipedon of Forces	27
Geometrical Composition and Resolution of Forces	28
Components in the Direction of two Axes	30
Components in the Direction of three Axes	32
Projection of Forces	34
Composition of a Group of Forces in a Plane	35
Composition of a Group of Forces in Space	36
Expression for the Resultant of two Forces	37
Principle of Moments	40
Principle of Virtual Moments	43

	PAGE
Resultant of Parallel Forces	45
Composition and Resolution—Parallel Forces	48
Lever arm of the Resultant	51
Centre of Parallel Forces	52
Resultant of a Group in a Plane	53
Tendency to Rotation—Equilibrium in a Plane	58
Equilibrium of Forces in Space	59
Equilibrium of a Revolving Body	60

CHAPTER III.

Weight—Centre of Gravity	62
Centre of Gravity of Straight Line	64
Of Symmetrical Lines and Areas	64
Of a Triangle	65
Of a Parallelogram—Of a Trapezoid	66
Of a Polygon	67
Of a Pyramid	68
Of Prisms, Cylinders, and Polyhedrons	70
Centre of Gravity Experimentally	71
Centre of Gravity by means of the Calculus	72
Centre of Gravity of an Arc of a Circle	73
Of a Parabolic Area	74
Of a Semi-Ellipsoid	75
Pressure and Stability	80
Problems in Construction	85

CHAPTER IV.

Definition of a Machine	94
Elementary Machines—Cord	96
The Lever	98
The Compound Lever	101
The Elbow-joint Press	102
The Balance	103

CONTENTS.

	PAGE
The Steelyard	105
The Bent Lever Balance—Compound Balances	106
The Inclined Plane	110
The Pulley	112
Single Pulley	113
Combinations of Pulleys	115
The Wheel and Axle	117
Combinations of Wheels and Axles	118
The Windlass	119
The Capstan—The Differential Windlass	120
Wheel-work	121
The Screw	123
The Differential Screw	125
Endless Screw	126
The Wedge	127
General Remarks on Machines	129
Friction	130
Limiting Angle of Resistance	133
Rolling Friction—Adhesion	135
Stiffness of Cords	136
Atmospheric Resistance—Friction on Inclined Planes	137
Line of least Fraction	140
Friction on Axle	141

CHAPTER V.

Uniform Motion	143
Varied Motion	144
Uniformly Varied Motion	146
Application to Falling Bodies	148
Bodies Projected Upwards	150
Restrained Vertical Motion	153
Atwood's Machines	156
Motion on Inclined Planes	158
Motion down a Succession of Inclined Planes	161
Periodic Motion	163

	PAGE
Angular Velocity	165
The Simple Pendulum	166
The Compound Pendulum	169
Practical Applications of the Pendulum	175
Graham's and Harrison's Pendulums	176
Basis of a System of Weights and Measures	177
Centre of Percussion	179
Moment of Inertia	180
Application of Calculus to Moment of Inertia	182
Centre of Gyration	186

CHAPTER VI.

Motion of Projectiles	188
Centripetal and Centrifugal Forces	197
Measure of Centrifugal Force	197
Centrifugal Force of Extended Masses	203
Principal Axes	206
Experimental Illustrations	207
Elevation of the outer rail of a Curved track	209
The Conical Pendulum	210
The Governor	212
Work	215
Work, when the Power acts obliquely	217
Work, when the Body moves on a Curve	219
Rotation—Quantity of Work	223
Accumulation of Work	225
Living Force of Revolving Bodies	227
Fly Wheels	228
Composition of Rotations	230
Application to Gyroscope	232

CHAPTER VII.

Classification of Fluids	236
Principle of Equal Pressures	236

CONTENTS.

	PAGE.
Pressure due to Weight	238
Centre of Pressure on a Plane Surface	243
Buoyant Effect of Fluids	249
Floating Bodies	249
Specific Gravity	251
Hydrostatic Balance	253
Specific Gravity of an Insoluble Body	253
Specific Gravity of Liquids	254
Specific Gravity of Soluble Bodies	255
Specific Gravity of Air and Gases	256
Hydrometers—Nicholson's Hydrometer	257
Scale Areometer	258
Volumeter	259
Densimeter	260
Centesimal Alcoholometer of Gay Lussac	261
Thermometer	263
Velocity of a Liquid through an Orifice	265
Spouting of Liquids on Horizontal Planes	268
Modifications due to Pressures	269
Coefficients of Efflux and Velocity	270
Efflux through short Tubes	272
Motion of Water in open Channels	274
Motion of Water in Pipes	277
General Remarks	279
Capillary Phenomena	280
Elevation and Depression between Plates	281
Attraction and Repulsion of Floating Bodies	282
Applications of the principle of Capillarity	283
Endosmose and Exosmose	284

CHAPTER VIII.

Gases and Vapors	285
Atmospheric Air	285
Atmospheric Pressure	286
Mariotte's Law	287

CONTENTS.

	PAGE
Gay Lussac's Law	290
Manometers—The open Manometer	291
The closed Manometer	292
The Siphon Guage	294
The Barometer—Siphon Barometer	295
The Cistern Barometer	296
Uses of the Barometer	297
Difference of Level	298
Work of Expanding Gas or Vapor	304
Efflux of a Gas or Vapor	306
Steam	308
Work of Steam	310
Experimental Formulas	311

CHAPTER IX.

Pumps—Sucking and Lifting Pumps	313
Sucking and Forcing Pump	318
Fire Engine	321
The Rotary Pump	322
Hydrostatic Press	324
The Siphon	326
Wurtemburg and Intermitting Siphon	328
Intermitting Springs	328
Siphon of Constant Flow—Hydraulic Ram	329
Archimedes' Screw	331
The Chain Pump—The Air Pump	332
Artificial Fountains—Hero's Ball	336
Hero's Fountain	337
Wine-Taster and Dropping Bottle	338
The Atmospheric Inkstand	338

MECHANICS.

CHAPTER I.

DEFINITIONS AND INTRODUCTORY REMARKS.

Definition of Natural Philosophy.

1. NATURAL PHILOSOPHY is that branch of Science which treats of the laws of the material universe.

These laws are called *laws of nature;* and it is assumed that they are *constant*, that is, that *like causes always produce like effects*. This principle, which is the basis of all Science, is an inductive truth founded upon universal experience.

Definition of a Body.

2. A BODY is a collection of material particles. When the dimensions of a body are exceedingly small, it is called a *material point*.

Rest and Motion.

3. A body is *at rest* when it retains the same *absolute* position in space; it is *in motion* when it continually changes its position.

A body is at rest with respect to surrounding objects, when it retains the same *relative* position with respect to them; it is in motion with respect to them, when it continually changes this relative position. These states are called *relative rest* and *relative motion*, to distinguish them from *absolute rest* and *absolute motion*. It is highly probable that no object in the universe is in a state of absolute rest.

Trajectory.

4. The path traced out, or described by a moving point, is called its *trajectory*. When this trajectory is a straight line, the motion is *rectilinear;* when it is a curved line, the motion is *curvilinear*.

Translation and Rotation.

5. When all of the points of a body move in parallel straight lines, the motion is called *motion of translation;* when the points of a body describe arcs of circles about a straight line, the motion is called *motion of rotation*. Other varieties of motion result from a combination of these two.

Uniform and Varied Motion.

6. The *velocity* of a moving point, is its rate of motion. When the point moves over equal spaces in any arbitrary equal portions of time, the motion is *uniform*, and the velocity is *constant;* when it moves over unequal spaces in equal portions of time, the motion is *varied*, and the velocity is *variable*. If the velocity continually increases, the motion is *accelerated;* if it continually decreases, the motion is *retarded*.

Forces.

7. A Force is anything which tends to change the state of a body with respect to rest or motion.

If a body is at rest, anything which tends to put it in motion is a force; if it is in motion, anything which tends to make it move faster, or slower, is a force. The power with which a force acts, is called its *intensity*.

Forces are of two kinds: *extraneous*, those which act upon a body from without; *molecular*, those which are exerted between adjacent particles of bodies.

An extraneous force may act for an instant and then cease, in which case it is called an *impulse*, or an *impulsive force;* or it may act continuously, in which case it is called an *incessant force*. An incessant force may be regarded as made up of a succession of impulses acting at equal but exceedingly small intervals of time. When these successive impulses are equal, the force is *constant;* when they are unequal, the force is *variable*. The force of gravity at any given place, is an example of a constant force; the effort of expanding steam, is an example of a variable force.

Molecular forces are of two kinds; *attractive*, those which tend to draw particles together; *repellent*, those which tend to separate them. These forces also exert an arranging power by virtue of which the particles of bodies are grouped into definite shapes. The phenomena of crystalization present examples of this action. Molecular forces of both kinds are continually exerted between the particles of all bodies, and upon their variation, in intensity and direction, depend the conditions of bodies, whether solid, liquid, or gaseous.

Classification of Bodies.

8. Bodies are divided into two classes, *solids* and *fluids*. A *solid* is a body which has a tendency to retain a permanent form. The particles of a solid adhere to each other so as to require the action of an extraneous force of greater or less intensity to separate them. A *fluid* is a body whose particles move freely amongst each other, each particle yielding to the slightest force. Fluids are divided into *liquids* and *gases*, liquids being sensibly incompressible, whilst gases are highly compressible. Many bodies are capable of existing in either of these states according to their temperature. Thus *ice, water,* and *steam,* are simply three different states of the same body.

Gravity.

9. Experiment and observation have shown that the earth exercises a force of attraction upon all bodies, tending to draw them towards its centre. This force, which is exerted upon every particle of every body, is called *the force of gravity*.

When a body is supported, the force of gravity produces pressure or *weight*; when it is unsupported, the force produces *motion*. Experiment and observation have shown that the entire force of attraction exerted by the earth upon any body, varies *directly as the quantity of matter in the body, and inversely as the square of its distance from the centre of the earth.* This force of attraction is mutual, so that the body attracts the earth according to the same law. Observation has shown that this law of mutual attraction extends throughout the universe, and for this reason it has received the name of *universal gravitation.*

Weight.

10. The WEIGHT of a body is the resultant action of the force of gravity upon all of its particles. If the body therefore remain the same, its weight at different places will vary directly as the force of gravity, or inversely as the square of its distance from the centre of the earth.

Mass.

11. The MASS of a body is the quantity of matter which it contains. Were the force of gravity the same at every point of the earth's surface, the weight of a body might be taken as the measure of its mass. But it is found that the force of gravity increases slightly in passing from the equator towards either pole, and consequently the weight of the same body increases as it is moved from the equator towards either pole; its *mass*, however, remains the same. If we take the weight of a body at the equator as the measure of its mass, it follows from what has just been said, that the *mass* will be equal to the weight at any place, divided by the force of gravity at that place, the force of gravity at the equator being regarded as the unit; or, denoting the mass of any body by M, its weight at any place by W, and the force of gravity at that place by g, we shall have

$$M = \frac{W}{g}; \text{ whence, } W = Mg.$$

The expression for the mass of a body is constant, as it should be, since the quantity of matter remains the same.

The UNIT OF MASS is any definite mass assumed as a standard of comparison. It may be one pound, one ounce, or any other unit of weight, taken at the equator. The pound is generally assumed as the unit of mass. The terms *weight* and *mass* may be regarded as synonymous, provided we understand that the weight is taken at the equator.

Density.

12. The DENSITY of a body is the quantity of matter contained in a unit of volume of the body, or it is the mass of a unit of volume.

At the same place the densities of two bodies are proportional to the weights of equal volumes. The mass of any body is therefore equal to its volume multiplied by its density, or denoting the volume by V, and the density by D, we have

$$M = VD.$$

We have also,

$$D = \frac{M}{V} = \frac{W}{Vg}; \text{ whence, } W = VDg.$$

Momentum.

13. The MOMENTUM of a moving body, or its QUANTITY of MOTION, is the product obtained by multiplying the mass moved, by the velocity with which it is moved; that is, we multiply the number of units in the mass moved by the number of units in the velocity with which it is moved and the product is the number of units in the momentum. This will be explained more in detail hereafter.

Properties of Bodies.

14. All bodies are endowed with certain attributes, or properties, the most important of which are, *magnitude* and *form; impenetrability; mobility; inertia; divisibility*, and *porosity; compressibility, dilatibility* and *elasticity; attraction, repulsion*, and *polarity*.

Magnitude and Form.

15. Magnitude is that property of a body by virtue of which it occupies a definite portion of space; every body possesses the three attributes of extension, length, breadth, and height. The form of a body is its figure or shape.

Impenetrability.

16. Impenetrability is that property by virtue of which no two bodies can occupy the same space at the same time. The particles of one body may be thrust aside by those of another, as when a nail is driven into wood; but where one body is, no other body can be.

Mobility.

17. Mobility is that property by virtue of which a body may be made to occupy different positions at different instants of time. Since a body cannot occupy two positions at the same instant, a certain interval must elapse whilst the body is passing from one position to another. Hence motion requires time, the idea of time being very closely connected with that of motion.

Inertia.

18. Inertia is that property by virtue of which a body tends to continue in the state of rest or motion in which it may be placed, until acted upon by some force. A body at rest cannot set itself in motion, nor can a body in motion increase or diminish its rate, or change the direction of its motion. Hence, *if a body is at rest, it will remain at rest, or if it is in motion, it will continue to move uniformly in a straight line, until acted upon by some force.* This principle is called the *law of inertia*. It follows immediately from this law, that if a force act upon a body in motion, it will impart the same velocity, and in the same general direction as though the body were at rest. It also follows that if a body, free to move, be acted upon simultaneously by two or more forces in the same, or in different directions, it will move in the general direction of each force, as though the other did not exist.

When a force acts upon a body at rest to produce motion, or upon a body in motion to change that motion, a resistance is developed equal and directly opposed to the effective force exerted. This resistance, due to inertia, is called the *force of inertia*. The effect of this resistance is called *re-action*, and the principle just explained may be expressed by saying that *action and re-action are equal and directly opposed*. This principle is called the *law of action and re-action*.

These two laws are deduced from observation and experiment, and upon them depends the mathematical theory of mechanics.

Divisibility and Porosity.

19. Divisibility is that property by virtue of which a body may be separated into parts. All bodies may be divided, and by successive divisions the fragments may be rendered very small. It is probable that all bodies are composed of ultimate atoms which are indivisible and indestructible; if so, they must be exceedingly minute. There are microscopic beings so small that millions of them do not equal in bulk a single grain of sand, and yet these animalcules possess organs, blood, and the like. How inconceivably minute, then, must be the atoms of which these various parts are composed.

Porosity is that property by virtue of which the particles of a body are more or less separated. The intermediate spaces are called *pores*. When the pores are small, the body is said to be *dense*; when they are large, it is said to be *rare*. Gold is a dense body, air or steam a rare one.

Compressibility, Dilatability, and Elasticity.

20. Compressibility, or contractility, is that property by virtue of which the particles of a body are susceptible of being brought nearer together, and dilatability is that property by virtue of which they may be separated to a greater distance. All bodies contract and expand when their temperatures are changed. Atmospheric air is an example of a body which readily contracts and expands.

Elasticity is that property by virtue of which a body tends to resume its original form after compression, or extension. Steel and India rubber are instances of elastic bodies. No bodies are perfectly elastic, nor are any perfectly inelastic. The force which a body exerts in endeavoring to resume its form after distortion, is called the *force of restitution*. If we denote the force of distortion by d, the force of restitution by r, and their ratio by e, we shall have

$$e = \frac{r}{d},$$

in which e is called the *modulus of elasticity*. Those bodies are most elastic which give the greatest value for e. Glass is highly elastic, clay is very inelastic.

Attraction, Repulsion, and Polarity.

21. Attraction is that property by virtue of which one particle has a tendency to pull others towards it. Repulsion is that property by virtue of which one particle tends to push others from it. The dissimilar poles of two magnets attract each other, whilst similar poles repel each other. It is supposed that forces of attraction and repulsion are continually exerted between the neighboring particles of bodies, and that the positions of these particles are continually changing, as these forces vary.

Polarity is that property by virtue of which the attractive and repellent forces between the particles exert an arranging power, so as to give definite forms to masses. The phenomena of crystalization already referred to, depend upon this property. It is to polarity that many of the most interesting phenomena of physics are to be attributed.

Equilibrium.

22. A system of forces is said to be *in equilibrium* when they mutually counteract each other's effects. If a system of forces in equilibrium be applied to a body, they will not change its state with respect to rest or motion; if the body be at rest it will remain so, or if it be in motion, it will continue to move uniformly, so far as these forces are concerned. The idea of an equilibrium of forces does not imply either rest or motion, but simply a continuance in the previous state, with respect to rest or motion. Hence two kinds of equilibrium are recognized; the equilibrium of rest, called *statical equilibrium*, and the equilibrium of motion, called *dynamical equilibrium*. If we observe that a body remains at rest, we infer that all the forces acting upon it are in equilibrium; if we observe that a body moves uniformly, we in like manner infer that all the forces acting upon it are in equilibrium.

Definition of Mechanics.

23. MECHANICS is that science which treats of the laws of equilibrium and motion. That branch of it which treats of the laws of equilibrium is called *statics*; that branch which treats of the laws of motion is called *dynamics*. When the bodies considered are liquids, of which water is a type, these two branches are called *hydrostatics* and *hydrodynamics*. When the bodies considered are gases, of which air is a type, these branches are called *ærostatics* and *ærodynamics*.

Measure of Forces.

24. We know nothing of the absolute nature of forces, and can only judge of them by their effects. We may, however, compare these effects, and in so doing, we virtually compare the forces themselves. Forces may act to produce pressure, or to produce motion. In the former case, they are called *forces of pressure;* in the latter case, *moving forces*. There are two corresponding methods of measuring forces, *first*, by the pressure they can exert, *secondly*, by the quantities of motion which they can communicate.

A force of pressure may be expressed in pounds; thus, a pressure of one pound is a force which, if directed vertically upwards, would just sustain a weight of one pound; a pressure of two pounds is a force which would sustain a weight of two pounds, and so on.

A moving force may be a single impulse, or it may be made up of a succession of impulses.

The *unit* of an impulsive force, is an impulse which can cause a unit of mass to move over a unit of space in a unit of time. A force which can cause two units of mass to move over a unit of space in a unit of time, or which can cause a unit of mass to move over two units of space in a unit of time, is called a double force.

A force which can cause three units of mass to move over a unit of space in a unit of time, or which can cause a unit of mass to move over three units of space in a unit of time, is called a *triple force*, and so on.

If we represent a unit of force by 1, a double force will be represented by 2, a triple force by 3, and so on.

In general, a force which can cause m units of mass to move over n units of space in a unit of time, will be represented by $m \times n$. Hence, forces may be compared with each other as readily as numbers, and by the same general rules.

The unit of mass, the unit of space, and the unit of time, are altogether arbitrary, but having been once assumed they must remain the same throughout the same discussion. We shall assume a mass weighing *one pound* at the equator, as the unit of mass, *one foot*, as the unit of space, and *one second*, as the unit of time.

Let us denote any impulsive force, by f, the mass moved, by m, and the velocity which the impulse can impart to it by v. Then, since the velocity is the space passed over in one second, we shall have, from what precedes,

$$f = mv.$$

If we suppose m to be equal to 1, we shall have,

$$f = v.$$

That is, *the measure of an impulse is the velocity which it can impart to a unit of mass.*

An incessant force is made of a succession of impulses. It has been agreed to take, as the measure of an incessant force, the quantity of motion that it can generate in one second, or the unit of time.

If we denote an incessant force by f, the mass moved by m, and the velocity generated in one second by v, we shall have,

$$f = mv.$$

If we suppose m to be equal to 1, we shall have,

$$f = v.$$

That is, *the measure of an incessant force is the velocity which it can generate in a unit of mass in a unit of time.*

If the force is of such a nature as to act equally upon every particle of a body, as gravity, for instance, the velocity generated will be entirely independent of the mass. In these cases, the velocity that a force can generate in a unit of time, is called the *acceleration* due to the force. If we denote the acceleration by f, the mass acted upon by m, and the entire moving force by f', we shall have,

$$f' = mf = mv.$$

Since an incessant force is made up of a succession of impulses, its measure may be assimilated to that of an impulsive force, so that both may be represented and treated in the same manner.

Forces of pressure, if not counteracted, would produce motion; and, as they differ in no other respect from the forces already considered, they also may be assimilated to impulsive forces, and treated in the same manner.

Representation of Forces.

25. It has been found convenient in Mechanics to represent forces by straight lines; this is readily effected by taking lines proportional to the forces which they represent. Having assumed some definite straight line to represent a unit of force, a double force will be represented by a line twice as long, a triple force by a line three times as long, and so on.

A force is completely given when we have its *intensity*, its *point of application*, and the *direction* in which it acts. When a force is represented by a straight line, the length of the line represents the *intensity*, one extremity of the line represents the *point of application*, and the direction of the line represents the *direction* of the force.

Fig. 1.

Thus, in figure 1, OP represents the *intensity*, O the *point of application*, and the direction from O to P is the *direction* of the force. This direction is generally indicated by an arrow head. It is to be observed that the point of application of a force may be taken at any point of its line of direction, and it is often found convenient to transfer it from one point to another on this line.

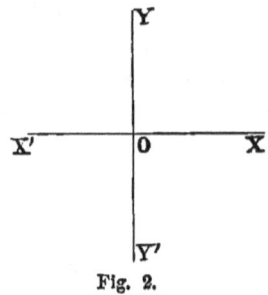

The intensity of a force may be represented analytically by a letter, which letter is usually the one placed at the arrow head; thus, in the example just given, we should designate the force OP by the single letter P.

If forces acting in any direction are regarded as positive, those acting in a contrary direction must be regarded as negative. This convention enables us to apply the ordinary rules of analysis to the investigations of Mechanics.

Forces situated in the same plane are generally referred to two rectangular axes, OX and OY, which are called *co-ordinate axes*. The direction from O towards X is that of *positive* abscissas; that from O towards X' is that of *negative* abscissas. The directions from O towards Y and Y', respectively, are those of *positive* and *negative* ordinates. Forces acting in the directions of positive abscissas and positive ordinates are positive; those acting in contrary directions, are negative.

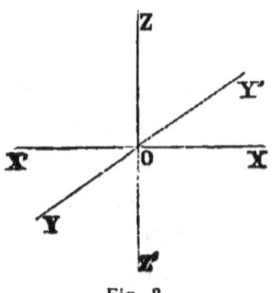

Forces in space are referred to three rectangular co-ordinate axes, OX, OY, and OZ. Forces acting from O towards X, Y, or Z, are *positive*, those acting in contrary directions, are *negative*.

CHAPTER II.

COMPOSITION, RESOLUTION, AND EQUILIBRIUM OF FORCES.

Composition of Forces whose directions coincide.

26. Composition of forces, is the operation of finding a single force whose effect is equivalent to that of two or more given forces. This single force is called the *resultant* of the given forces. Resolution of forces, is the operation of finding two or more forces whose united effect is equivalent to that of a given force. These forces are called *components* of the given force.

If two forces are applied at the same point, and act in the same direction, their resultant is equal to the sum of the two forces. If they act in contrary directions, their resultant is equal to their difference, and acts in the direction of the greater one. In general, if any number of forces are applied at the same point, some of which act in one direction, and the others in a contrary direction, their resultant is equal to the sum of those which act in one direction, diminished by that of those which act in the contrary direction · or, if we regard the rule for signs, the resultant is equal to the *algebraic sum of the components ;* the sign of this algebraic sum makes known the direction in which the resultant acts. This principle follows immediately from the rule adopted for measuring forces.

Thus, if the forces P, P', &c., applied at any point, act in the direction of positive abscissas, whilst the forces P'', P''', &c., applied to the same point, act in the direction of negative abscissas, then will their resultant, denoted by R, be given by the equation,

$$R = (P + P' + \&c.,) - (P'' + P''' + \&c.)$$

If the first term of the second member of this equation is numerically greater than the second, R is positive, which shows that the resultant acts in the direction of positive abscissas. If the first term is numerically less than the second, R is negative, which shows that the resultant acts in the direction of negative abscissas.

If the two terms of the second member are numerically equal, R will reduce to 0. In this case, the forces will exactly counterbalance each other, and, consequently, will be in equilibrium.

Whenever a system of forces is in equilibrium, their resultant must necessarily be equal to 0. When all of the forces of the system are applied at the same point, this single condition will be sufficient to determine an equilibrium.

All of the forces of a system which act in the general direction of the same straight line, are called *homologous*, and their *algebraic sum* may be expressed by writing the expression for a single force, prefixing the symbol Σ, a symbol which indicates the *algebraic sum of several homologous* quantities. We might, for example, write the preceding equation under the form,

$$R = \Sigma (P) \quad \ldots \quad (1.)$$

This equation expresses the fact, that the *resultant of a system of forces, acting in the same direction, is equal to the algebraic sum of the forces.*

Parallelogram of Forces.

27. Let P and Q be two forces applied to the material point O, taken as a unit of mass, and acting in the directions OP and OQ. Let OP represent the velocity generated by the force P, and OQ the velocity generated by the force Q. Draw PR parallel to OQ, and QR parallel to OP; draw also the diagonal OR.

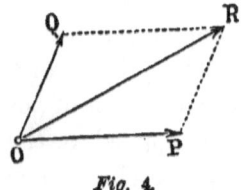

Fig. 4.

From the law of inertia (Art 18), it follows that a mass acted upon by two simultaneous forces moves in the general direction of each, as though the other did not exist. Now, if we suppose the material point O, to be acted upon simultaneously by the two forces P and Q, it will, by virtue of the first, be found at the end of one second somewhere on the line PR; and by virtue of the second somewhere on the line QR; hence, it will be at their point of intersection. But had the point O been acted upon by a single force, represented in direction and intensity by OR, it would have moved from O to R in the same time. Hence, the single force R is equivalent, in effect, to the aggregate of the two forces P and Q; it is, therefore, their resultant. Hence,

If two forces be represented in direction and intensity by the adjacent sides of a parallelogram, their resultant will be represented in direction and intensity by that diagonal of the parallelogram which passes through their point of intersection.

This principle is called the *parallelogram of forces.*

In the preceding demonstration we have only considered *moving forces*, but the principle is equally true for forces of pressure; for, if we suppose a force equal and directly opposed to the resultant R, this force will be in equilibrium with the forces P and Q, which will then become forces of pressure. The relation between the forces will not be changed by this hypothesis, and we may therefore enunciate the principle as follows:

If two pressures be represented in direction and intensity by the adjacent sides of a parallelogram, their resultant will be represented in direction and intensity by that diagonal of the parallelogram which passes through their common point.

This principle is called the *parallelogram of pressures.*

Hence, we see that *moving forces* and *pressures* may be compounded and resolved according to the same principles, and by the same general laws.

Parallelopipedon of Forces.

28. Let P, Q, and S represent three forces applied to the same point, and not in the same plane. Upon these lines, as edges, construct the parallelopipedon OR, and draw OM and SR. From the preceding article, OM represents the resultant of P and Q, and from the same article, OR represents the resultant of OM and S.

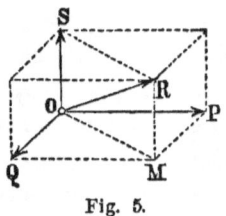

Fig. 5.

Hence, OR is the resultant of the three forces P, Q, and S. That is, *if three forces be represented in direction and intensity by three adjacent edges of a parallelopipedon, their resultant will be represented by that diagonal of the parallelopipedon which passes through their point of intersection.*

This principle is known as *the parallelopipedon of forces*, and is equally true for moving forces and pressures.

Geometrical Composition and Resolution of Forces.

29. The following constructions depend upon the principle of the parallelogram of forces.

1. Having given the directions and intensities of two forces applied at the same point, to find the direction and intensity of their resultant.

Let OP and OQ represent the given forces, and O their point of application; draw PR parallel to OQ, and QR parallel to OP, and draw the diagonal OR; it will be the resultant sought.

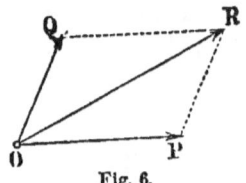

Fig. 6.

2. Having given the direction and intensity of the resultant of two forces, and the direction and intensity of one of its components, to find the direction and intensity of the other component.

Let R be the given resultant, P the given component, and O their point of application; draw RP, and through O draw OQ parallel to RP, also through R draw RQ parallel to PO; then will OQ be the component sought.

3. Having given the direction and intensity of the resultant of two forces, and the directions of the two components, to find the intensities of the components.

COMPOSITION AND RESOLUTION OF FORCES.

Let R be the given resultant, OP and OQ the directions of the components, and O their point of application. Through R draw RP and RQ respectively, parallel to QO and PO, then will OP and OQ represent the intensities of the components.

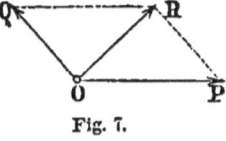

Fig. 7.

From this construction it is evident that any force may be resolved into two components having any direction whatever; these, again may each be resolved into new components, and so on; hence it follows that a single force may be resolved into any number of components having any assumed directions whatever.

4. Having given the direction and intensity of the resultant of two forces, and the intensities of the components, to find their directions.

Let R be the given resultant, and O its point of application. With R as a centre, and one of the components as a radius, describe an arc of a circle; with O as a centre, and the other component as a radius, describe a second arc cutting the first at P; draw PR and PO, and complete the parallelogram PQ, then will OP and OQ be the directions sought.

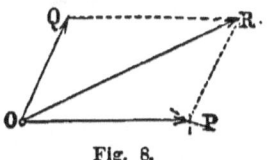

Fig. 8.

5. To find the resultant of any number of forces, P, Q, S, T, &c., lying in the same plane, and applied at the same point. Construct the resultant R' of P and Q, then construct the resultant R'' of R' and S, then the resultant R of R'' and T, and so on: the final resultant will be the resultant of the system.

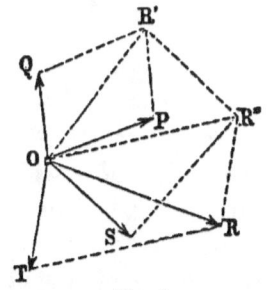

Fig. 9.

By inspecting the preceding figure, we see that in the polygon $OQR'R''RT$, the side QR' is equal and parallel to the force P, the side $R'R''$ to the force S, and the side $R''R$ to the force T, and so on. Hence, we may construct the resultant of such a system of forces by drawing through the second extremity of the first force, a line parallel and equal to the second force, through the second extremity of this line, a line parallel and equal to the third force, and so on to the last. The line drawn from the starting point to the last extremity of the last line drawn, will represent the resultant sought. If the last extremity of the last force fall at the starting point, the resultant will be 0, and the system will be in equilibrium.

This principle is called the *polygon of forces;* its simplest case is the *triangle of forces.*

Components of a Force in the direction of two axes.

30. To find expressions for the components of a force which act in directions parallel to two rectangular axes. Let OX and OY be two such axes, and R any force lying in their plane; construct the components parallel to OX and OY, as before explained, and denote the angle LAR, which the force makes with the axis of X, by α. From the figure, we have,

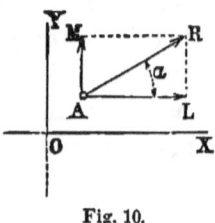

Fig. 10.

$$AL = R \cos \alpha, \text{ and } RL = AM = R \sin \alpha\,;$$

or, making $AL = X$, and $AM = Y$, we have,

$$X = R \cos \alpha, \text{ and } Y = R \sin \alpha \quad . \quad . \quad (2.)$$

The angle α is estimated from the direction of positive abscissas around to the left through 360°.

For all values of α from 0° to 90°, and from 270° to 360°, the cosine of α will be positive, and, consequently, the component AL will be positive; that is, it will act in the direction of positive abscissas. For all values of α from 90° to 270°, the cosine of α will be negative, and the component AL will act in the direction of negative abscissas.

For all values of α from 0° to 180°, the sine of α will be positive, and the component AM will be positive; that is, it will act in the direction of positive ordinates. For all values of α from 180° to 360°, the sine of α will be negative, and the component AM will act in the direction of negative ordinates.

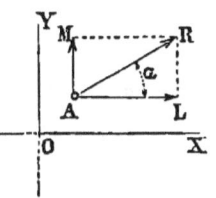

Fig. 10.

For α = 90°, or α = 270°, we shall have $AL = 0$. For α = 0, or α = 180°, we shall have $AM = 0$.

If we regard AL and AM as two given forces, R will be their resultant; and since $RL = AM$, we shall have from the figure,

$$R = \sqrt{X^2 + Y^2} \quad \ldots \quad (3.)$$

Hence, *the resultant of any two forces, at right-angles to each other, is equal to the square root of the sum of the squares of the two forces.*

From the figure, we also have,

$$\cos \alpha = \frac{X}{R}, \text{ and } \sin \alpha = \frac{Y}{R}.$$

Hence, the resultant is completely determined.

PRACTICAL EXAMPLES.

1. Two pressures of 9 and 12 pounds, respectively, act upon a point, and at right-angles to each other. Required, the direction and intensity of the resultant pressure.

SOLUTION.

We have,

$X = 9$, and $Y = 12$; ∴ $R = \sqrt{81 + 144} = 15$.

Also, $\cos \alpha = \dfrac{9}{15} = .6$; ∴ α = 53° 7′ 47″.

That is, the resultant pressure is 15 lbs., and it makes an angle of 53° 7′ 47″ with the direction of the first force.

2. Two forces are to each other as 3 is to 4, and their

resultant is 20 lbs. What are the intensities of the components?

SOLUTION.

We have, $3Y = 4X$, or $Y = \tfrac{4}{3}X$, and $R = 20$;

$$\therefore 20 = \sqrt{X^2 + \tfrac{16}{9}X^2} = \tfrac{5}{3}X;$$

Hence, $X = 12$, and $Y = 16$.

3. A boat fastened by a rope to a point on the shore, is urged by the wind perpendicular to the current, with a force of 18 pounds, and down the current by a force of 22 pounds. What is the tension, or strain, upon the rope, and what angle does it make with the current?

SOLUTION.

We have

$X = 22$, and $Y = 18$; $\quad\therefore R = \sqrt{808} = 28.425$;

Also, $\cos \alpha = \dfrac{22}{28.425}$; $\quad\therefore \alpha = 39° \; 17' \; 20''$.

Hence the tension is 28.425 lbs., and the angle $39° \; 17' \; 20''$.

Components of a Force in the direction of three axes.

31. To find expressions for the components of a force in the directions of three rectangular axes. Let OR represent the force, and OX, OY, and OZ, three rectangular axes drawn through its point of application, O. Construct a parallelopipedon on OR as a diagonal, having three of its edges coinciding with the axes. Then will the lines OL, OM, and ON, represent the required components. Denote these components, respectively, by X, Y, and Z. Draw lines from R, to L, M, and

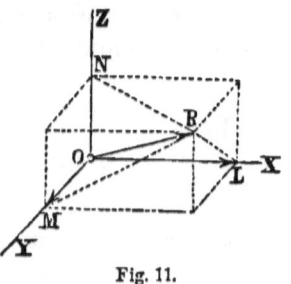

Fig. 11.

COMPOSITION AND RESOLUTION OF FORCES. 33

N, respectively; these will be perpendicular to the axes, and with them, and the force R, will form three right-angled triangles. Denote the angle between R and the axis of X by α, that between R and the axis of Y by β, and that between R and the axis of Z by γ; we shall have from the right-angled triangles referred to, the following equations:

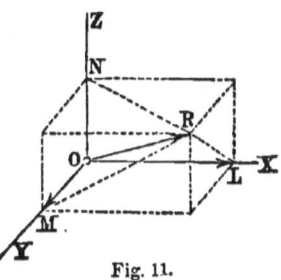

Fig. 11.

$$X = R \cos \alpha, \quad Y = R \cos \beta, \text{ and } Z = R \cos \gamma.$$

The angles α, β, and γ, are estimated from the directions of the positive co-ordinates, through 360°. The components above found will be positive when they act in the direction of positive co-ordinates, and negative when they act in a contrary direction.

If we regard X, Y, and Z, as three forces, R will be their resultant, and we shall have, from a known property of the rectangular parallelopipedon,

$$R = \sqrt{X^2 + Y^2 + Z^2} \quad \ldots \quad (4.)$$

That is, *the resultant of three forces at right angles to each other, is equal to the square root of the sum of the squares of the components.*

We also have from the figure,

$$\cos \alpha = \frac{X}{R}, \cos \beta = \frac{Y}{R}, \text{ and } \cos \gamma = \frac{Z}{R},$$

Hence, the position of the resultant is completely determined.

EXAMPLES.

1. Required the intensity and direction of the resultant of three forces at right angles to each other, having the intensities 4, 5, and 6 pounds, respectively.

SOLUTION

We have,

$X = 4$, $Y = 5$, and $Z = 6$. $\therefore R = \sqrt{77} = 8.775$.

Also, $\cos \alpha = \dfrac{4}{8.775}$, $\cos \beta = \dfrac{5}{8.775}$, and $\cos \gamma = \dfrac{6}{8.775}$; whence, $\alpha = 62°52'51''$, $\beta = 55°15'50''$, and $\gamma = 46°51'43''$.

Hence the resultant pressure is 8.775 lbs., and it makes, with the components taken in order, angles equal to 62° 52' 51'', 55° 15' 50'', and 46° 51' 43''.

2. Three forces at right angles are to each other as the numbers 2, 3, and 4, and their resultant is 60 lbs. What are the intensities of the forces?

SOLUTION.

We have

$$Y = \tfrac{3}{2}X, \quad Z = 2X, \text{ and } R = 60;$$

Hence,

$$60 = \sqrt{X^2 + \tfrac{9}{4}X^2 + 4X^2} = \tfrac{1}{2}X\sqrt{29} = 2.6925 X.$$

$$\therefore X = 22.284.$$

The components are, therefore,

22.284 lbs., 33.426 lbs., and 44.568 lbs.

Projection of Forces.

32. If planes be passed through the extremities of a force, perpendicular to the direction of any straight line, that portion of the line intercepted between them is the *projection of the force upon the line*. The operation of resolving forces into components in the direction of rectangular axes, is nothing more than that of finding their projections upon these axes.

If two straight lines be drawn through the extremities of a force, perpendicular to any plane, and the points in which they meet the plane be joined by a straight line, this line is *the projection of the force upon the plane*.

If we denote any force by P, and the angle which it makes with any line or plane by α, $P\cos\alpha$ will represent the projection of the force on the line or plane. In both cases the projection of the force is its *effective component* in the direction of the line or plane upon which it is projected.

Composition of a Group of Forces in a Plane.

33. Let P, P', P'', &c., denote any number of forces lying in the same plane, and applied at a common point, and represent the angles which they make with the axis of X by $\alpha, \alpha', \alpha''$, &c. Their components in the direction of the axis of X are $P\cos\alpha$, $P'\cos\alpha'$, $P''\cos\alpha''$, &c., and their components in the direction of the axis of Y, are $P\sin\alpha$, $P'\sin\alpha'$, $P''\sin\alpha''$, &c.

If we denote the resultant of the group of components which are parallel to the axis of X by X, and the resultant of the group parallel to the axis of Y by Y, we shall have, (Art. 26),

$$X = \Sigma\,(P\cos\alpha), \text{ and } Y = \Sigma\,(P\sin\alpha) \quad . \quad . \quad (5.)$$

The resultant of X and Y is the same as the resultant of the given forces. Denoting this resultant by R, and recollecting that X and Y are perpendicular to each other, we have, as in Article 30,

$$R = \sqrt{X^2 + Y^2} \quad . \quad . \quad . \quad . \quad (6.)$$

If we denote the angle which the resultant makes with the axis of X by a, we shall have, as in Article 30,

$$\cos a = \frac{X}{R}, \text{ and } \sin a = \frac{Y}{R}.$$

EXAMPLES.

1. Three forces, whose intensities are respectively equal to 50, 40, and 70, lie in the same plane, and are applied at the same point, and make with an axis through that point, angles equal to 15°, 30°, and 45°, respectively. Required the intensity and direction of the resultant.

SOLUTION.

We have,

$X = 50 \cos 15° + 40 \cos 30° + 70 \cos 45° = 132.435$,

and

$Y = 50 \sin 15° + 40 \sin 30° + 70 \sin 45° = 82.44$;

whence,

$$R = \sqrt{6798 + 17539} = 156.$$

and $\quad \cos a = \dfrac{132.435}{156}; \quad \therefore\ a = 31° 54' 12''.$

The resultant is 156, and the angle which it makes with the axis is equal to 31° 54' 12''.

2. Three forces 4, 5, and 6, lie in the same plane, making equal angles with each other. Required the intensity of their resultant and the angle which it makes with the least force.

SOLUTION.

Take the least force as the axis of X. Then the angle between it and the second force is 120°, and that between it and the third force is 240°. We have

$$X = 4 + 5 \cos 120° + 6 \cos 240° = -1.5;$$
$$Y = 5 \sin 120° + 6 \sin 240° = -.866;$$
$$\therefore R = \sqrt{3},\ \cos a = -\frac{1.5}{1.732},\ \sin a = -\frac{.866}{1.732};$$
$$\therefore a = 210°.$$

3. Two forces, one of 5 lbs. and the other of 7 lbs., are applied at the same point, and make with each other an angle of 120°. What is the intensity of their resultant?

Ans. 6.24 lbs.

Composition of a Group of Forces in Space.

34. Let the forces be represented by P, P', P'', &c. The angles which they make with the axis of X, by α, α', α'', &c., the angles which they make with the axis of Y, by β, β', β'', &c., and the angles which they make with the axis of Z by $\gamma, \gamma', \gamma''$, &c. Resolving each force into components, respectively parallel to the three co-ordinate axes, and denoting the resultants of the groups in the directions of the respective axes by X, Y, and Z, we shall have, as in the preceding article,

$$X = \Sigma\,(P \cos \alpha), \quad Y = \Sigma\,(P \cos \beta), \quad Z = \Sigma\,(P \cos \gamma.)$$

If we denote the resultant of the system by R, and the angles which it makes with the axes by a, b, and c, we shall have, as in Article 31,

$$R = \sqrt{X^2 + Y^2 + Z^2}.$$

$$\cos a = \frac{X}{R}, \quad \cos b = \frac{Y}{R}, \quad \text{and } \cos c = \frac{Z}{R}.$$

The application of these formulas is entirely analogous to that of the formulas in the preceding article.

Expression for the Resultant of two Forces.

35. Let us consider two forces, P and P', situated in the same plane. Since the position of the co-ordinate axes is perfectly arbitrary, let the axis of X be so taken as to coincide with the force P; α will then be equal to 0, and we shall have $\sin \alpha = 0$, and $\cos \alpha = 1$. The value of X (Equation 5), will become $P + P' \cos \alpha'$, and the value of Y will become $P' \sin \alpha'$. Squaring these values, substituting them in Equation (6), and reducing by the relation $\sin^2 \alpha' + \cos^2 \alpha' = 1$, we have,

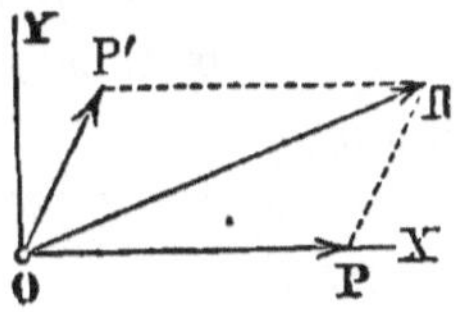

Fig. 12.

$$R = \sqrt{P^2 + P'^2 + 2PP' \cos \alpha'} \quad . \quad (7.)$$

The angle α' is the angle included between the given forces. Hence,

The resultant of any two forces, applied at the same point, is equal to the square root of the sum of the squares

of the two forces, plus twice the product of the forces into the cosine of their included angle.

If we make α' greater than 90°, and less than 270°, its cosine will be negative, and we shall have,

$$R = \sqrt{P^2 + P'^2 - 2PP' \cos \alpha'}.$$

If we make $\alpha' = 0$, its cosine will be 1, and we shall have,

$$R = P + P'.$$

If we make $\alpha' = 90°$, its cosine will be equal to 0, and we shall have,

$$R = \sqrt{P^2 + P'^2}.$$

If we make $\alpha' = 180°$, its cosine will be -1, and we shall have,

$$R = P - P'.$$

The last three results conform to principles already deduced. Let P and Q be two forces, and R their resultant. The figure QP being a parallelogram, the side PR is equal to Q. From the triangle ORP we have, in accordance with the principles of trigonometry,

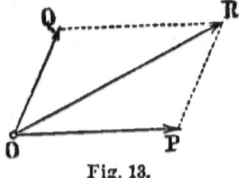

Fig. 13.

$$P : Q : R :: \sin ORP : \sin ROP : \sin OPR. \quad (8.)$$

If we apply a force R' equal and directly opposed to R, the forces P, Q, and R', will be in equilibrium. The angles ORP, and QOR', being opposite exterior and interior angles, are supplements of each other; hence, $\sin ORP = \sin QOR'$. The angles ROP, and POR', are adjacent, and, consequently, supplementary; hence, $\sin ROP = \sin POR'$. The angles

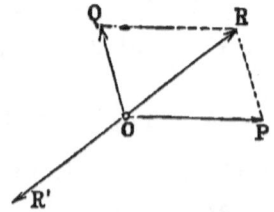

Fig 14.

OPR, and POQ, are interior angles on the same side, and, consequently, supplementary; hence, $\sin OPR = \sin POQ$. We have also $R = R'$. Making these substitutions in the preceding proportion, we have,

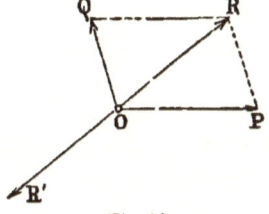

Fig 14.

$$P : Q : R' :: \sin QOR' : \sin POR' : \sin POQ.$$

Hence, *if three forces are in equilibrium, each is proportional to the sine of the angle between the other two.*

EXAMPLES.

1. Two forces, P and Q, are equal in intensity to 24 and 30, respectively, and the angle between them is 105°. What is the intensity of their resultant?

$$R = \sqrt{24^2 + 30^2 + 2 \times 24 \times 30 \cos 105°} = 33.21.$$

2. Two forces, P and Q, whose intensities are respectively equal to 5 and 12, have a resultant whose intensity is 13. Required the angle between them.

$$13 = \sqrt{25 + 144 + 2 \times 5 \times 12 \cos \alpha}.$$

$$\therefore \cos \alpha = 0, \text{ or } \alpha = 90°. \quad Ans.$$

3. A boat is impelled by the current at the rate of 4 miles per hour, and by the wind at the rate of 7 miles per hour. What will be her rate per hour when the direction of the wind makes an angle of 45° with that of the current?

$$R = \sqrt{16 + 49 + 2 \times 4 \times 7 \cos 45°} = 10.2\text{m}. \quad Ans.$$

4. A weight of 50 lbs., suspended by a string, is drawn aside by a horizontal force until the string makes an angle of 30° with the vertical. Required the value of the horizontal force, and the tension of the string.

Ans. 28.8675 lbs., and 57.735 lbs.

5. Two forces, and their resultant, are all equal. What is the value of the angle between the two forces? 120°.

6. A point is kept at rest by three forces of 6, 8, and 11 lbs., respectively. Required the angles which they make with each other.

SOLUTION.

We have $P = 8$, $Q = 6$, and $R' = 11$. Since the forces are in equilibrium, we shall have $R' = R = 11$; hence from the preceding article,

$$11 = \sqrt{64 + 36 + 96 \cos QOP};$$

$$\therefore \cos QOP = \tfrac{21}{96}; \text{ or, } QOP = 77° \, 21' \, 52''.$$

From the last proportion we have,

$$\frac{\sin POR'}{\sin QOP} = \frac{6}{11}; \quad \therefore \sin POR' = .53224;$$

or, $$POR' = 147° \, 50' \, 34''.$$

Also, $$\frac{\sin QOR'}{\sin QOP} = \frac{8}{11}; \quad \therefore \sin QOR' = .70965;$$

or, $$QOR' = 134° \, 47' \, 34''$$

Principle of Moments.

36. The moment of a force, with respect to a point, is the product obtained by multiplying the intensity of the force by the perpendicular distance from the point to the line of direction of the force.

The fixed point is called the *centre of moments;* the perpendicular distance is called the *lever arm of the force*; and the moment itself measures the tendency of the force to produce rotation about the centre of moments.

COMPOSITION AND RESOLUTION OF FORCES. 41

Let P and Q be any two forces, and R their resultant; assume any point C, in their plane, as the centre of moments, and from it, let fall upon the directions of the forces, the perpendiculars, Cp, Cq, and Cr; denote these perpendiculars respectively by p, q, and r.

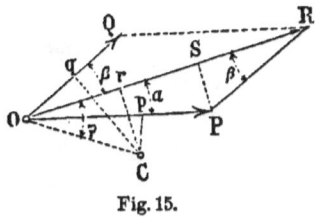

Fig. 15.

Then will Pp, Qq, and Rr, be the moments of the forces P, Q, and R. Draw CO, and from P let fall the perpendicular PS, upon OR. Denote the angle ROP, by α, the angle ROQ, or its equal, ORP, by β, and the angle ROC by φ.

Since $PR = Q$, we have from the right-angled triangles OPS and PRS, the equations,

$$R = Q \cos \beta + P \cos \alpha.$$
$$0 = Q \sin \beta - P \sin \alpha.$$

Multiplying both members of the first equation by $\sin \varphi$, and both members of the second by $\cos \varphi$, then adding the resulting equations, we find,

$$R \sin \varphi = Q\ (\sin \varphi \cos \beta + \sin \beta \cos \varphi) + P\ (\sin \varphi \cos \alpha - \sin \alpha \cos \varphi).$$

Whence, by reduction,

$$R \sin \varphi = Q \sin (\varphi + \beta) + P \sin (\varphi - \alpha).$$

From the figure, we have,

$$\sin \varphi = \frac{r}{OC}, \quad \sin (\varphi - \alpha) = \frac{p}{OC}, \quad \text{and} \sin (\varphi + \beta) = \frac{q}{OC}.$$

Substituting in the preceding equation, and reducing, we have,

$$Rr = Qq + Pp.$$

When the point C falls within the angle POR, $\varphi - \alpha$ becomes negative, and the equation just deduced becomes

$$Rr = Qq - Pp.$$

Hence, we conclude in all cases, that *the moment of the resultant of two forces is equal to the algebraic sum of the moments of the forces taken separately.*

If we regard the force Q as the resultant of two others, and one of these in turn, as the resultant of two others, and so on, the principle may be extended to any number of forces lying in the same plane, and applied at the same point. This principle may, in the general case, be expressed by the equation

$$Rr = \Sigma\,(Pp) \quad \ldots \ldots \quad (9.)$$

That is, *the moment of the resultant of any number of forces, lying in the same plane, and applied at the same point, is equal to the algebraic sum of the moments of the forces taken separately.*

This is called *the principle of moments.*

The moment of the resultant is called the *resultant moment;* the moments of the components are called *component moments;* and the plane passing through the resultant and centre of moments, is the *plane of moments.*

When a force tends to turn its point of application about the centre of moments, in the direction of the motion of the hands of a watch, its moment is considered positive; consequently, when it tends to produce rotation in a contrary direction, the moment must be negative. If the resultant moment is negative, the tendency of the system is to produce rotation in a negative direction about the centre of moments. If the resultant moment is 0, there is no tendency to produce rotation in the system. The resultant moment may become 0, either in consequence of the lever arm becoming 0, or in consequence of the resultant itself being equal to 0. In the former case, the centre of moments lies upon the direction of the resultant, and the numerical value of the sum of the moments of the forces which tend to produce rotation in one direction, is equal to that of those which tend to produce motion in a contrary direction. In the latter case, the system of forces is in equilibrium.

Moments, with respect to an Axis.

37. To form an idea of the moment of a force with respect to a straight line, taken as an *axis of moments*. Let P represent any force, and let the axis of Z be assumed so as to coincide with the axis of moments. Draw the straight line AB perpendicular, both to the direction of the force and to the axis of moments; at the point A, in which this perpendicular

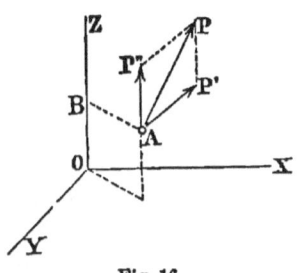

Fig. 16.

intersects the direction of the force, let the force P be resolved into two components, P'' and P', the first parallel to the axis of Z, and the second at right angles to it. The former will have no tendency to produce rotation, the latter will tend to produce rotation, which tendency will be measured by $P' \times AB$; this product is the moment of the force P with respect to the axis of moments, and is evidently equal to the moment of the projection of the force upon a plane at right-angles to the axis, taken with respect to the point in which this axis pierces the plane as a centre of moments.

If there are any number of forces situated in any manner in space, it is clear from the preceding principles that *their resultant moment, with respect to any straight line taken as an axis of moments, is equal to the algebraic sum of the component moments with respect to the same axis.*

Principle of Virtual Moments.

38. Let P represent a force applied to the material point O; let the point O be moved by an extraneous force to some position, C, very near to O; project the path OC upon the direction of the force; the projection Op, or Op', is called the

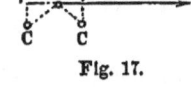

Fig. 17.

virtual velocity of the force, and is taken *positively* when it falls upon the direction of the force, as Op, and *nega-*

tively when it falls upon the prolongation of the force, as *Op'*. The product obtained by multiplying any force by its virtual velocity is called the *virtual moment* of the force

Assume the figure and notation of Article 36. *Op*, *Oq*, and *Or* are the virtual velocities of the forces *P*, *Q*, and *R*. Let us denote the virtual velocity of any force by the symbol of variation δ, followed by a small letter of the same name as that which designates the force.

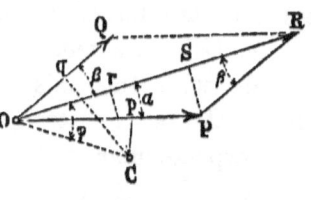

Fig. 18.

We have from the figure, as in Article 36, the relations,

$$R = P \cos \alpha + Q \cos \beta.$$

$$0 = P \sin \alpha - Q \sin \beta.$$

Multiplying both members of the first by cos φ, and of the second by sin φ, and adding the resultant equations, we have,

$$R \cos \varphi = P (\cos \alpha \cos \varphi + \sin \alpha \sin \varphi) + Q (\cos \varphi \cos \beta - \sin \varphi \sin \beta).$$

Or, by reduction,

$$R \cos \varphi = P \cos (\varphi - \alpha) + Q \cos (\varphi + \beta).$$

But, from the right-angled triangles *COp*, *COq*, and *COr*, we have,

$$\cos \varphi = \frac{\delta r}{OC}, \cos (\varphi - \alpha) = \frac{\delta p}{OC}, \text{ and } \cos (\varphi + \beta) = \frac{\delta q}{OC};$$

Substituting these in the preceding equation, and reducing, we have,

$$R \delta r = P \delta p + Q \delta q.$$

Hence, *the virtual moment of the resultant of two forces, is equal to the algebraic sum of the virtual moments of the two forces taken separately.*

If we regard the force Q as the resultant of two other forces, and one of these as the resultant of two others, and so on, the principle may be extended to any number of forces, applied at the same point. This principle may be expressed by the following equation:

$$R\delta r = \Sigma\, (P\delta p) \quad \ldots \quad (10.)$$

Hence, *the virtual moment of the resultant of any number of forces applied at the same point, is equal to the algebraic sum of the virtual moments of the forces taken separately.*

This is called *the principle of virtual moments.* If the resultant is equal to 0, the system is in equilibrium, and the algebraic sum of the virtual moments is equal to 0; conversely, if the algebraic sum of the virtual moments of the forces is equal to 0, the resultant is also equal to 0, and the forces are in equilibrium.

This principle, and the preceding one, are much used in discussing the subject of machines.

Resultant of parallel Forces.

39. Let P and Q be two forces lying in the same plane, and applied at points invariably connected, for example, at the points M and N of a solid body. Their lines of direction being prolonged, will meet at some point O; and if we suppose the points of application to be transferred to O, their resultant may be determined by the parallelogram of forces. The direction of the resultant will pass through O. (Art. 27.) Whether the forces be transferred to O or not, the direction of the resultant will always pass through O, and this whatever may be the value of the included angle. Now, supposing the points of application to be at M and N, let the force Q be turned about N as an axis. As it approaches parallelism with P,

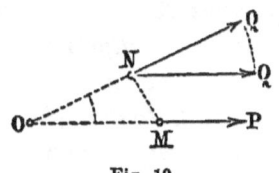

Fig. 19.

the point O will recede from M and N, and the resultant will also approach parallelism with P. Finally, when Q becomes parallel to P, the point O will be at an infinite distance from M and N, and *the resultant will also be parallel to P and Q.* In any position of P and Q, the value of the resultant, denoted by R, will be given by the equation (Art. 36),

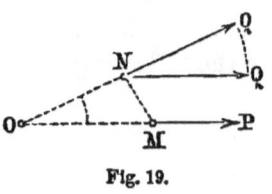

Fig. 19.

$$R = P\cos\alpha + Q\cos\beta.$$

When the forces are parallel, and lying in the same direction, we shall have $\alpha = 0$, and $\beta = 0$; or, $\cos\alpha = 1$, and $\cos\beta = 1$. Hence,

$$R = P + Q.$$

If the forces lie in opposite directions, we shall have $\alpha = 0$, and $\beta = 180°$; or, $\cos\alpha = 1$, and $\cos\beta = -1$. Hence,

$$R = P - Q.$$

That is, *the resultant of two parallel forces is equal in intensity to the algebraic sum of the forces, and its line of direction is parallel to that of the two forces.*

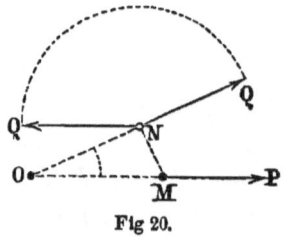

Fig 20.

If we regard Q as the resultant of two parallel forces, and one of these as the resultant of two others, and so on, the principle may be extended to any number of parallel forces. Denoting the resultant of a group of parallel forces, P, P', P'', &c., by R, we have,

$$R = \Sigma(P) \quad . \quad . \quad . \quad . \quad (11.)$$

That is, *the resultant of a group of parallel forces is equal in intensity to the algebraic sum of the forces. Its line of direction is also parallel to that of the given forces.*

COMPOSITION AND RESOLUTION OF FORCES.   47

Point of Application of the Resultant.

40. Let P and Q be two parallel forces, and R their resultant. Let M and N be the points of application of the two forces, and S the point in which the direction of R cuts the line MN. Through N draw NL perpendicular to the general direction of the forces, and assume the point C, in which it intersects the line of direction of R, as a centre of moments. Since the centre of moments is on the line of direction of the resultant, the lever arm of the resultant will be 0, and we shall have, from the principle of moments (Art. 36),

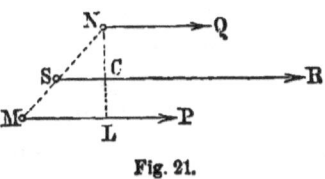

Fig. 21.

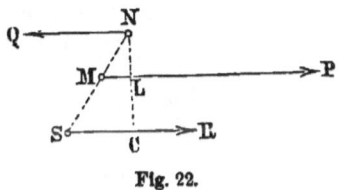

Fig. 22.

$$P \times CL = Q \times CN;$$

or, $$P : Q :: CN : CL.$$

But, from the similar triangles CNS and LNM, we have,

$$CN : CL :: SN : SM.$$

Combining the two proportions, we have,

$$P : Q :: SN : SM.$$

That is, *the line of direction of the resultant divides the line joining the points of application of the components, inversely as the components.*

From the last proportion, we have, by composition,

$$P : Q : P + Q :: SN : SM : SN + SM;$$

and, by division,

$$P : Q : P - Q :: SN \cdot SM : SN - SM.$$

When the forces act in the same direction, $P + Q$ will be their resultant, and $SN + SM$ will equal MN. Since $P + Q$ is greater than either P or Q, MN will be greater than either SN or SM, which shows that the resultant lies between the components.

When the forces act in contrary directions, $P - Q$ will be their resultant, and $SN - SM$ will equal MN. Since $P - Q$ is less than P (supposed the greater of the components), MN will be less than SN, which shows that the resultant lies without both components, and on the side of the greater.

Substituting in the preceding proportions, for $P + Q$, $P - Q$, $SN + SM$, and $SN - SM$, their values, we have,

$$P : Q : R : : SN : SM : MN \ldots (8)'.$$

That is, *of two parallel forces and their resultant, each is proportional to the distance between the other two.*

Geometrical Composition and Resolution of Parallel Forces.

41. The preceding principles give rise to the following geometrical constructions:

1. To find the resultant of two parallel forces lying in the same direction:

Let P and Q be the forces, M and N their points of application. Make $MQ' = Q$, and $NP' = P$; draw $P'Q'$, cutting MN in S; through S draw SR parallel to MP, and make it equal to $P + Q$: it will be the resultant.

For, from the similar triangles $P'SN$ and $Q'SM$, we have,

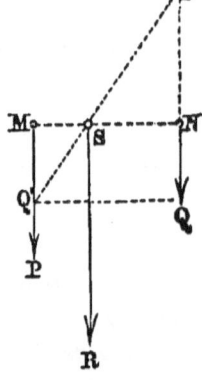

Fig. 23.

$$P'N : Q'M : : SN : SM; \text{ or, } P : Q : : SN : SM.$$

After the construction is made, the distances MS and NS may be measured by a scale of equal parts.

EXAMPLE.

Given $P = 9$ lbs, $Q = 6$ lbs., and $MN = 30$ in. Required MS.

We have $R = 15$, hence,

$15 : 6 : : 30 : MS$; $\therefore MS = 12$ in. **Ans.**

2. To find the resultant of two parallel forces acting in opposite directions:

Let P and Q be the forces, M and N their points of application. Prolong QN till $NA = P$, and make $MB = Q$; draw AB, and produce it till it cuts NM produced in S; draw SR parallel to MP, and make it equal to BP, it will be the resultant required.

For from the similar triangles SNA and SMB, we have,

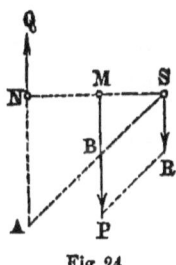

Fig. 24.

$AN : BM : : SN : SM$; or, $P : Q : : SN : SM$.

EXAMPLE.

Given $P = 20$ lbs., $Q = 8$ lbs., and $NM = 18$ in. Required SN.

We have $R = 20 - 8 = 12$; hence, from Proportion (8),

$12 : 20 : : 18 : SN$; $\therefore SN = 30$ in. **Ans.**

3. To resolve a given force into two parallel components lying in the same direction, and applied at given points:

Let R be the given force, M and N the given points of application. Through M and N draw lines parallel to R. Make $MA = R$, and draw AN, cutting R in B; make $MP = SB$ and $NQ = BR$; they will be the required components.

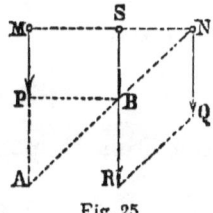

Fig. 25.

For, from the similar triangles AMN and BSN,

$$BS : AM :: SN : MN;$$
or, $$BS : R :: SN : MN.$$

But, from Proportion (8)', we have,

$$P : R :: SN : MN;$$

$\therefore$ $BS = P$, and $BR = Q$.

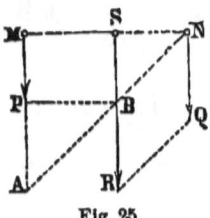

Fig. 25.

EXAMPLE.

Given $R = 24$ lbs., $SM = 7$ in., and $SN = 5$ in. Required P and Q.

From Proportion (8), we have,

$$12 : 7 :: 24 : Q; \quad \therefore \quad Q = 14 \text{ lbs.}$$
$$12 : 5 :: 24 : P; \quad \therefore \quad P = 10 \text{ lbs.}$$

4. To resolve a given force into parallel components lying in opposite directions, and applied at given points. Both points of application must lie on the same side of the given force. Let R be the given force, M and N the given points of application. Through M and N draw lines parallel to R; make $NB = R$, and draw BM; through S, draw SA parallel to MB; then will NA and BA be equal to the intensities of the components. Make $MP = AN$, and $NQ = AB$, and they will be the components. For, from the triangles ASN, and BMN, we have,

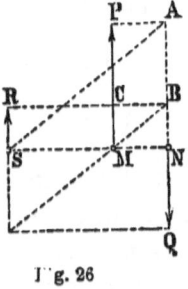

Fig. 26.

$$AN : BN :: SN : MN; \text{ or, } AN : R :: SN : MN.$$

But, from Proportion (8)', we have,

$$P : R :: SN : MN; \quad \therefore \quad AN = P, \text{ and } AB = Q.$$

COMPOSITION AND RESOLUTION OF FORCES. 51

EXAMPLE.

Given $R = 24$ lbs., $SN = 18$ in., and $SM = 9$ in. Required P and Q.

From Proportion (8)', we have,

$P : 24 : : 18 : 9;$ $\therefore$ $P = 48$ lbs.
$Q : 24 : : 9 : 9;$ $\therefore$ $Q = 24$ lbs.

$$R = P - Q = 24 \text{ lbs.}$$

5. To find the resultant of any number of parallel forces.

Let P, P', P'', P''', be such a system of forces. Find the resultant of P and P', by the rule already given, it will be $R' = P + P'$; find the resultant of R' and P'', it will be $R'' = P + P' + P''$; find the resultant of R'' and P''', it will be $R = P + P' + P'' + P'''$. If there is a greater number of forces, the operation of composition may be continued; the final result will be the resultant of the system. If some of the

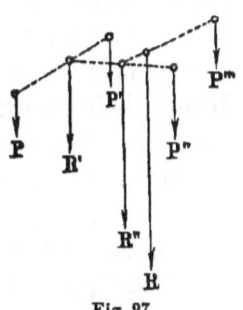

Fig. 27.

forces act in contrary directions, combine all which act in one direction, as just explained, and call their resultant R'; then combine all those which act in a contrary direction, and call their resultant R''; finally, combine R' and R'' by a preceding rule; their resultant R will be the resultant of the system.

If $R' = R''$, the resultant will be 0, and its point of application will be at an infinite distance. In this case, the forces reduce to a *couple*, the effect of which is simply to produce rotation.

Lever Arm of the Resultant.

42. Let P, P', P'', &c., denote any number of parallel forces, and p, p', p'', &c., their lever arms with respect to an axis of moments, taken perpendicular to the common direction of the forces; denote the lever arm of the resultant of the system, taken with respect to the same axis, by r From the principle of moments (Art. 37),

$$(P + P' + P'' + \&c.)r = Pp + P'p' + \&c.;$$

or, $$r = \frac{\Sigma(Pp)}{\Sigma(P)} \quad \ldots \quad (12.)$$

Hence, *the lever arm of the resultant of a system of parallel forces, with respect to an axis at right-angles to their direction, is equal to the algebraic sum of the moments of the forces divided by the algebraic sum of the forces.*

Centre of Parallel Forces.

43. Let there be any number of forces, P, P', P'', &c., applied at points invariably connected together, and whose co-ordinates are x, y, z; x', y', z'; x'', y'', z''; &c. Let R denote their resultant, and represent the co-ordinates of its point of application, by $x_1, y_1,$ and z_1; denote the angles made by the common direction of the forces with the axes of X, Y, and Z, by α, β, and γ.

Suppose each force resolved into three components, respectively parallel to the co-ordinate axes, the points of application being unchanged:

The components parallel to the axis of X are,

$$P\cos\alpha, \quad P'\cos\alpha, \quad P''\cos\alpha, \quad \&c., \quad R\cos\alpha;$$

those parallel to the axis of Y are,

$$P\cos\beta, \quad P'\cos\beta, \quad P''\cos\beta, \quad \&c., \quad R\cos\beta;$$

and those parallel to the axis of Z are,

$$P\cos\gamma, \quad P'\cos\gamma, \quad P''\cos\gamma, \quad \&c., \quad R\cos\gamma.$$

If we take the moments of the components parallel to the axis of Z, with respect to the axis of Y, as an axis of moments, we shall have, for the lever arms of the components, $x, x', x'',$ &c.; and from the principle of moments (Art. 37),

$$R\cos\gamma \, x_1 = P\cos\gamma \, x + P'\cos\gamma \, x' + \&c.$$

Striking out the common factor cos γ, and substituting for R its value, we have,

$$\Sigma(P)x_1 = \Sigma(Px);$$

whence, $$x_1 = \frac{\Sigma(Px)}{\Sigma(P)}.$$

In like manner, if we take the moments of the same components, with respect to the axis of X, we shall have,

$$y_1 = \frac{\Sigma(Py)}{\Sigma(P)}.$$

And, if we take the moments of the components parallel to the axis of Y, with respect to the axis of X, we shall have,

$$z_1 = \frac{\Sigma(Pz)}{\Sigma(P)}.$$

Hence we have for the co-ordinates of the point of application of the resultant,

$$x_1 = \frac{\Sigma(Px)}{\Sigma(P)}, \quad y_1 = \frac{\Sigma(Py)}{\Sigma(P)}, \text{ and } z_1 = \frac{\Sigma(Pz)}{\Sigma(P)}. \quad (13.)$$

These co-ordinates are entirely independent of the direction of the parallel forces, and will remain the same so long as their intensities and points of application remain unchanged.

The point whose co-ordinates we have just found, is called the *centre of parallel forces*.

Resultant of a Group of Forces in a Plane, and applied at points invariably connected.

44. Let P, P', P'', &c., be any number of forces lying in the same plane, and applied at points invariably connected together; that is, at points of the same solid body.

Through any point O in the plane of the forces, draw any two straight lines, OX and OY, at right angles to each other, and lying in the plane of the forces; assume these as co-ordinate axes. Denote the angles which the forces P, P', P'', &c., make with the axis OX, by α, α', α'', &c., and the angles which they make with the axis OY, by β, β', β'', &c.; denote, also, the co-ordinates of the points of application of the forces, by x, y; x', y'; x'', y''; &c.

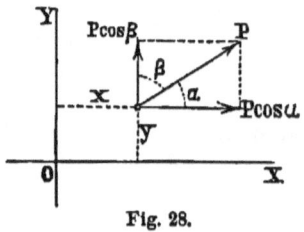

Fig. 28.

Let each force be resolved into components parallel to the co-ordinate axes; we shall have for the group parallel to the axis of X,

$$P\cos\alpha, \quad P'\cos\alpha', \quad P''\cos\alpha'', \text{ &c.};$$

and, for the group parallel to the axis of Y,

$$P\cos\beta, \quad P'\cos\beta', \quad P''\cos\beta'', \text{ &c.};$$

The resultant of the first group is equal to the algebraic sum of the components (Art. 39); denoting this by X, we shall have,

$$X = \Sigma(P\cos\alpha) \quad \ldots \quad (14.)$$

In like manner, denoting the resultant of the second group by Y, we shall have,

$$Y = \Sigma(P\cos\beta) \quad \ldots \quad (15.)$$

The forces X and Y intersect in a point, which is the point of application of the system of forces. Denoting the resultant by R, we shall have (Art. 33),

$$R = \sqrt{X^2 + Y^2}.$$

To find the point of application of R, let O be taken as a centre of moments, and denote the lever arms of X and Y

by y_1 and x_1, respectively. From the principle of Article 42, we shall have,

$$x_1 = \frac{\Sigma(P\cos\beta\, x)}{\Sigma(P\cos\beta)} \quad \ldots \quad (16.)$$

$$y_1 = \frac{\Sigma(P\cos\alpha\, y)}{\Sigma(P\cos\alpha)} \quad \ldots \quad (17.)$$

It we denote the angles which the resultant makes with the axes of X and Y by a and b respectively, we shall have, as in Article 33,

$$\cos a = \frac{X}{R}, \quad \cos b = \frac{Y}{R}. \quad \ldots \quad (18.)$$

Equations (16) and (17) make known the point of application, and Equations (18) make known its direction; hence, the resultant is completely determined.

To find the moment of R, with respect to O as a centre of moments, let us denote its lever arm by r, and the lever arms of P, P', P'', &c., with respect to O, by p, p', p'', &c.

The moment of the force $P\cos\alpha$, is $P\cos\alpha\, y$, and that of the force $P\cos\beta$, is $-P\cos\beta\, x$. The negative sign is given to the last result, because the forces $P\cos\alpha$ and $P\cos\beta$ tend to turn the system in contrary directions.

From the principle of moments (Art. 36), the moment of P is equal to the algebraic sum of the moments of its components. Hence,

$$Pp = P\cos\alpha\, y - P\cos\beta\, x.$$

In like manner, the moments of the other component forces may be found. Because the moment of the resultant is equal to the algebraic sum of the moments of all its components (Art. 36), we have,

$$Rr = \Sigma(Pp) = \Sigma(P\cos\alpha\, y - P\cos\beta\, x) \quad . \quad (19.)$$

Resultant of a Group of Forces situated in Space, and applied at points invariably connected.

45. Let P, P', P'', &c., be any number of forces situated in any manner in space, and applied at points of the same solid body. Assume any point O in space, and through it draw any three lines perpendicular to each other. Assume these lines as axes. Denote the angles which the forces P, P', P'', &c., make with the axis of X, by α, α', α'', &c.; the angles which they make with the axis

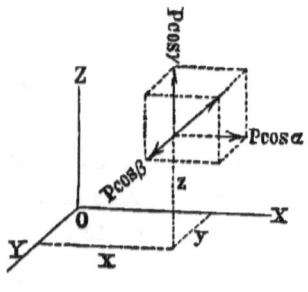

Fig. 29.

of Y, by β, β', β'', &c.; the angles which they make with the axis of Z, by γ, γ', γ'', &c., and denote the co-ordinates of their points of application by x, y, z; x', y', z'; x'', y'', z''; &c.

Let each force be resolved into components respectively parallel to the co-ordinate axes.

We shall have for the group parallel to the axis of X,

$$P\cos\alpha, \quad P'\cos\alpha', \quad P''\cos\alpha'', \quad \&c.;$$

for the group parallel to the axis of Y,

$$P\cos\beta, \quad P'\cos\beta', \quad P''\cos\beta'', \quad \&c.;$$

and for the group parallel to the axis of Z,

$$P\cos\gamma, \quad P'\cos\gamma', \quad P''\cos\gamma'', \quad \&c.$$

Denoting the resultants of these several groups by X, Y, and Z, we shall have,

$$X = \Sigma(P\cos\alpha,) \quad Y = \Sigma(P\cos\beta,) \quad \text{and} \quad Z = \Sigma(P\cos\gamma) \quad . \quad (20.)$$

If these three forces intersect at a point, this point is the point of application of the resultant of the entire system. Denote this resultant by R; then, since the forces X, Y, and Z are perpendicular to each other, we shall have,

$$R = \sqrt{X^2 + Y^2 + Z^2} \quad \ldots \quad (21.)$$

To find the Co-ordinates of the point of application of R.

Consider each of the forces, X, Y, and Z, with respect to the axis whose name comes next in order, and denote the lever arm of X, with respect to the axis of Y, by z_1; that of Y, with respect to the axis of Z, by x_1; and that of Z, with respect to the axis of X, by y_1. We shall have as in the last article,

$$\left. \begin{array}{l} x_1 = \dfrac{\Sigma(P\cos\beta\, x)}{\Sigma(P\cos\beta)} \\[2mm] y_1 = \dfrac{\Sigma(P\cos\gamma\, y)}{\Sigma(P\cos\gamma)} \\[2mm] z_1 = \dfrac{\Sigma(P\cos\alpha\, z)}{\Sigma(P\cos\alpha)} \end{array} \right\} \ \ldots \ (22.)$$

in which x_1, y_1, and z_1, are the co-ordinates of the point of application of R.

Denoting the angles which R makes with the axes by a, b, and c, respectively, we have, as in the preceding article,

$$\cos a = \frac{X}{R}, \ \cos b = \frac{Y}{R}, \ \cos c = \frac{Z}{R} \ . \ . \ (23.)$$

The values of X, Y, and Z, may be computed by means of Equations (20), and these being substituted in (21), make known the value of the resultant. The co-ordinates of its point of application result from Equations (22), and its line of direction is shown by Equations (23). The intensity, direction, and point of application being known, the resultant is completely determined.

Measure of the tendency to Rotation about the Axes.

46. Let X, Y, and Z denote the components of the resultant of the system, as in the last article, and denote, as before, the co-ordinates of the point of application of the resultant by x_1, y_1, and z_1. To find the resultant moment, with respect to the axis of Z, it may be observed that the component Z, can produce no rotary effect, since it is parallel to the axis of Z; the moment of the component Y, with respect to the axis of Z, is Yx_1; the moment of the component X, with respect to the same axis, is $-Xy_1$, the negative sign being taken because the force X tends to produce rotation in a negative direction. Hence, the resultant moment of the system, with respect to the axis of Z, is,

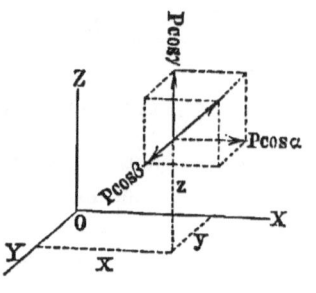

Fig. 30.

$$Yx_1 - Xy_1;$$

or, substituting for X and Y their values, we have,

$$Yx_1 - Xy_1 = \Sigma(P\cos\beta\, x - P\cos\alpha\, y) \quad . \quad (24.)$$

In like manner for the resultant moment of the system, with respect to the axis X,

$$Zy_1 - Yz_1 = \Sigma(P\cos\gamma\, y - P\cos\beta\, z) \quad . \quad (25.)$$

And for the resultant moment, with respect to the axis of Y,

$$Xz_1 - Zx_1 = \Sigma(P\cos\alpha\, z - P\cos\gamma\, x) \quad . \quad (26.)$$

Equilibrium of Forces in a Plane.

47. In order that a system of forces lying in the same plane, and applied at points of a free solid, may be in equilibrium, two conditions must be fulfilled: First, the resultant of the system must have no tendency to produce

EQUILIBRIUM OF FORCES. 59

motion of translation; and, secondly, it must have nc tendency to produce motion of rotation. Conversely, if these conditions are satisfied, the system will be in equilibrium.

The first condition will be fulfilled, and will only be fulfilled, when the resultant is equal to 0; but from Art. 44, we have,

$$R = \sqrt{X^2 + Y^2}.$$

The value of R can only be equal to 0 when $X = 0$, and $Y = 0$; or, what is the same thing,

$$\Sigma(P\cos\alpha) = 0, \text{ and } \Sigma(P\cos\beta) = 0 \quad . \quad (27.)$$

The second condition will be fulfilled, and will only be fulfilled, when the moment of the resultant, with respect to any point of the plane, is equal to 0, whence,

$$Rr = 0; \text{ or, } \Sigma(Pp) = 0 \quad . \quad . \quad . \quad (28.)$$

Hence, from Equations (27) and (28), in order that a system of forces, lying in the same plane, and applied at points of a free solid body, may be in equilibrium, we must have,

1st. *The algebraic sum of the components of the forces in the direction of any two rectangular axes separately equal to 0.*

2d. *The algebraic sum of the moments of the forces, with respect to any point in the plane, equal to 0.*

Equilibrium of Forces in Space.

48. In order that a system of forces situated in any manner in space, and applied at points of a free solid body, may be in equilibrium, two conditions must be fulfilled. First, the forces must have no tendency to produce motion of translation; and secondly, they must have no tendency to produce motion of rotation about either of the three rectangular axes. Conversely, when these conditions are fulfilled, the system will be in equilibrium. The first condition will be fulfilled, and will only be fulfilled, when the resultant is equal to 0. But, from Equation (21),

$$R = \sqrt{X^2 + Y^2 + Z^2}.$$

That this value of R may be 0, we must have, separately,

$$X = 0, \quad Y = 0, \text{ and } Z = 0;$$

or, what is the same thing,

$$\Sigma(P\cos\alpha) = 0, \quad \Sigma(P\cos\beta) = 0, \text{ and } \Sigma(P\cos\gamma) = 0 \quad . \quad (29.)$$

The second condition will be fulfilled, and will only be fulfilled, when the moments, with respect to each of the three axes, are separately equal to 0. This gives (Art. 46),

$$\left.\begin{array}{l}\Sigma(P\cos\beta\, x - P\cos\alpha\, y) = 0 \\ \Sigma(P\cos\gamma\, y - P\cos\beta\, z) = 0 \\ \Sigma(P\cos\alpha\, z - P\cos\gamma\, x) = 0\end{array}\right\} \cdot \cdot (30.)$$

Hence (Equations 29 and 30), in order that a system of forces in space applied at points of a free solid may be in equilibrium:

1st. *The algebraic sum of the components of the forces in the direction of any three rectangular axes must be separately equal to* 0.

2d. *The algebraic sum of the moments of the forces, with respect to any three rectangular axes, must be separately equal to* 0.

Equilibrium of Forces applied to a Revolving Body.

49. If a body is restrained by a fixed axis, about which it is free to revolve, we may take this line as the axis of X. Since the axis is fixed, there can be no motion of translation, neither can there be any rotation about either of the other two axes of co-ordinates. All of Equations (29), and the first and third of Equations (30), will be satisfied by virtue of the connection of the body with the fixed axis

The second of Equations (30) is, therefore, the only one that must be satisfied by the relation between the forces. We must have, therefore,

$$\Sigma(P\cos\gamma\, y - P\cos\beta\, z) = 0 \quad . \quad . \quad (31.)$$

That is, if a body is restrained by a fixed axis, the forces applied to it will be in equilibrium when *the algebraic sum of the moments of the forces with respect to this axis is equal to* 0.

CHAPTER III.

CENTRE OF GRAVITY AND STABILITY.

Weight.

50. That force by virtue of which a body, when abandoned to itself, falls towards the earth, is called the *force of gravity*. The force of gravity acts upon every particle of a body, and, if resisted, gives rise to a *pressure;* this pressure is called the *weight of the particle*. The resultant weight of all the particles of a body is called *the weight of the body*. The weights of the particles are sensibly directed towards the centre of the earth; but this point being nearly 4,000 miles from the surface, we may, for all practical purposes, regard these weights as parallel forces; hence, the weight of a body acts in the same direction as the weights of its elementary particles, and is equal to their sum.

Centre of Gravity.

51. The centre of gravity of a body is *the point of application of its weight*. The weight being the resultant of a system of parallel forces, the centre of gravity is a centre of parallel forces, and so long as the relative position of the particles remains unchanged, this point will retain a fixed position in the body, and this independently of any particular position of the body (Art. 43). The position of the centre of gravity is entirely independent of the value of the force of gravity, provided that we regard this force as constant throughout the dimensions of the body, which we may do in all practical cases. Hence, the centre of gravity is the same for the same body, wherever it may be situated. The determination of the centre of gravity is, then, reduced to the determination of the centre of a system of parallel forces. Equations (13) are, therefore, immediately appliplicable.

Preliminary discussion.

52. Let there be any number of weights applied at points of a straight line. We may take the axis of X to coincide with this line, and because the points of application of the weights are on this line, we shall have,

$$y = 0, \ y' = 0, \ \&c.; \quad z = 0, \ z' = 0, \ \&c.;$$

substituting these in the second and third of Equations (13), we have,

$$y_1 = 0, \text{ and } z_1 = 0.$$

Hence, *the point of application of the resultant is on the given line.*

In the case of a *material straight line*, that is, of a line made up of material points, the weight of each point will be applied at that point, and from what has just been shown, the point of application of the resultant weight will also be on the line; but this point is the centre of gravity of the line.

Hence, *the centre of gravity of a material straight line is situated somewhere on the line.*

Let weights be applied at points of a given plane. We may take the plane XY to coincide with this plane, and in this case we shall have,

$$z = 0, \ z' = 0, \ \&c.;$$

these in the third of Equations (13) will give,

$$z_1 = 0;$$

hence, *the point of application of the resultant weights is in the plane.*

It may be shown, as before, that *the centre of gravity of a material plane curve, or of a material plane area, is in the plane of the curve, or area.*

If the bodies considered are homogeneous in structure, the weights of any elementary portions are proportional to their volumes, and the problem for finding the centre of gravity is reduced to that for finding the centre of figure. In what follows, lines and surfaces will be considered as made up of material points, and all the volumes considered will be regarded as homogeneous unless the contrary is stated.

Centre of Gravity of a straight line.

53. Let there be two material points M and N, equal in weight, and firmly connected by an inflexible line MN. The resultant of these weights will bisect the line MN in S (Art. 40); hence S is the centre of gravity of the two points M and N.

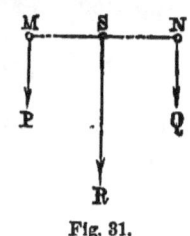
Fig. 31.

Let MN be a material straight line, and S its middle point. We may regard it as composed of heavy material points A, A'; B, B', &c., equal in weight, and so disposed that for each point on one side of S, there is another point on the other side of it and equally distant from it. From what precedes, the

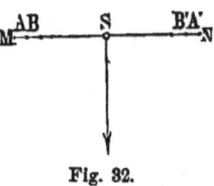
Fig. 32.

centre of gravity of each pair of points is at S, and consequently the centre of gravity of the whole line is at S. That is, *the centre of gravity of a straight line is at its middle point.*

Centre of Gravity of symmetrical lines and areas.

54. Let $APBQ$ be a plane curve, and AB a diameter, that is, a line which bisects a system of parallel chords; let PQ be one of the chords bisected. The centre of gravity of the chord PQ will be upon AB, and in like manner, the centre of gravity of any pair of points lying at the extremity of one of the parallel chords will be found upon the diam-

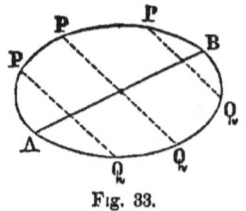
Fig. 33.

eter; hence, the centre of gravity of the entire curve is upon the diameter (Art. 52). The entire area of the curve is made up of the system of parallel chords bisected, and since the centre of gravity of each chord is upon the diameter, it follows that the centre of gravity of the area is upon the diameter.

Hence, *if any curve, or area, has a diameter, the centre of gravity of the curve, or area, lies upon that diameter.*

If a curve or area has two diameters, the centre of gravity will be found at their point of intersection. Hence, in the circle and ellipse the centre of gravity is at the centre of the curve.

If a surface has a diametral plane, that is, a plane which bisects a system of parallel chords terminating in the surface, then will the centre of gravity of the extremities of each chord lie in the diametral plane, and consequently, the centre of gravity of the surface will be in that plane. The centre of gravity of the volume bounded by such a surface, for like reason, lies in the diametral plane.

Hence, *if a surface, or volume, has a diametral plane, the centre of gravity of the surface, or volume, lies in that plane.* If a surface, or volume, has three diametral planes intersecting each other in a point, that point is the centre of gravity. Hence, the centre of gravity of the sphere and the ellipsoid lie at their centres. We see, also, that the centre of gravity of a surface, or volume, of revolution lies in the axis of revolution.

Centre of Gravity of a Triangle.

55. Let ABC be any plane triangle. Join the vertex A with the middle point D of the opposite side BC; then will AD bisect all of the lines drawn in the triangle parallel to the base BC; hence, the centre of gravity of the triangle lies upon AD (Art. 54); for a like reason, the centre of gravity of the triangle lies upon the line BE, drawn from the vertex B to the middle point of the opposite side AC; it is, therefore, at G, their point of intersection.

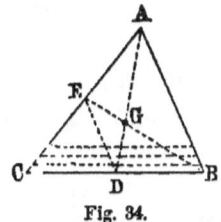

Fig. 34.

Draw ED; then, since ED bisects AC and BC, it is parallel to AB, and the triangles EGD and AGB are similar. The side ED is equal to one-half of its homologous side AB, consequently the side GD is equal to one-half of its homologous side AG; that is, the point G is one-third of the distance from D to A.

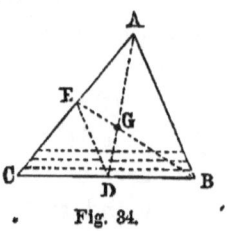

Fig. 34.

Hence, *the centre of gravity of a plane triangle is on a line drawn from the vertex to the middle point of the base, and at one-third of the distance from the base to the vertex.*

Centre of Gravity of a Parallelogram.

56. Let AC be any parallelogram. Draw EF bisecting the sides AB and CD; it will also bisect all lines of the parallelogram parallel to these sides; hence, the centre of gravity lies on it; draw also the line OH bisecting the sides AD and BC; for a similar reason, the centre of gravity lies on it: it is, therefore, at G, their point of intersection.

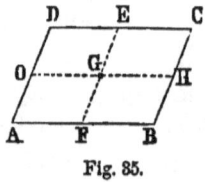

Fig. 35.

Hence, *the centre of gravity of a parallelogram lies at the point of intersection of two straight lines joining the middle points of the opposite sides.*

It is to be remarked, that this point coincides with the point of intersection of the diagonals of the parallelogram.

Centre of Gravity of a Trapezoid.

57. Let AC be a trapezoid. Join the middle points, O and P, of the parallel sides, by a straight line; this line will bisect all lines parallel to AB and DC; hence, it must contain the centre of gravity. Draw the diagonal BD, dividing the trapezoid into two triangles. Draw also the lines DO and BP; take

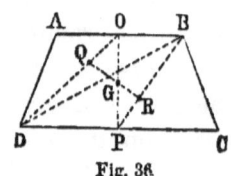

Fig. 36.

$OQ = \frac{1}{3}OD$, and $PR = \frac{1}{3}PB$; then will Q and R be the centres of gravity of these triangles (Art. 55). Join Q and R by a straight line; the centre of gravity of the trapezoid must be on this line (Art. 52). Hence, it is at G where the line QR cuts OP.

Centre of Gravity of a Polygon.

58. Let $ABCDE$ be any polygon, and a, b, c, d, e, the middle points of its sides. The weights of the sides will be proportional to their lengths, and may be represented by them. Let it first be required to find the centre of gravity of the perimeter; join a and b, and find a point o, such that

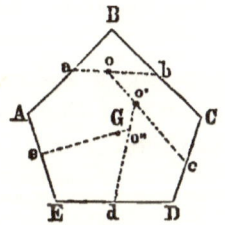

Fig. 37.

$$ao : ob :: BC : BA ;$$

then will o be the centre of gravity of the sides AB and BC. Join o and c, and find a point o', such that

$$oo' : o'c :: CD : AB + BC;$$

then will o' be the centre of gravity of the three sides, AB, BC, and CD. Join o' with d, and proceed as before, continuing the operation till the last point, G, is found; this will be the centre of gravity of the perimeter.

To find the centre of gravity of the area, divide it into the least number of triangles possible, and find the centre of gravity of each triangle. The weights of these triangles will be proportional to their areas, and may be represented by them. (Art. 52.) Let $ABCDEA$ be any polygon, and O, O', O'', the centres of gravity of the triangles into which it can be divided. Join O and O', and find a point O''', such that

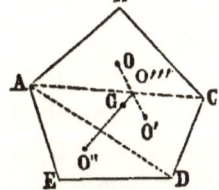

Fig. 38.

$$O'O''' : OO''' :: ABC : ACD;$$

then will O''' be the centre of gravity of the two triangles ABC and ACD.

Join O'' and O''', and find a point G, such that

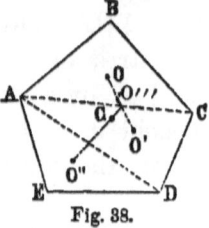

Fig. 38.

$$O'''G : O''G :: ADE : ABC + ACD;$$

then will G be the centre of gravity of the given polygon.

Every curvilinear area may be regarded as polygonal, the number of sides being very great. Hence, the centres of gravity of their perimeters and areas may be found by the methods given.

Centre of Gravity of a Pyramid.

59. Any triangular pyramid may be regarded as made up of infinitely thin layers parallel to either of its faces. If a straight line be drawn from either vertex to the centre of gravity of the opposite face, it will pass through the centres of gravity of all the layers parallel to that face. We may regard the weight of each layer as being applied at its centre of gravity, that is, at a point of this line; hence, the centre of gravity of the pyramid is on this line (Art. 52).

Let $ABCD$ be a pyramid, and K the middle point of DC. Draw KB and KA, and lay off $KO = \tfrac{1}{3}KB$, and $KO' = \tfrac{1}{3}KA$. Then will O be the centre of gravity of the face DBC, and O' that of the face CAD. Draw AO and BO' intersecting in G. Because the centre of gravity of the pyramid is upon both AO and BO', it is at their intersection G. Draw OO'; then KO and KO' being respectively third parts of KB and KA, OO' is parallel to AB, and the triangles OGO' and AGB are similar, consequently

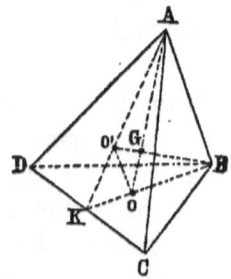

Fig. 39.

their homologous sides are proportional. But OO' is one-third of AB, consequently OG is one-third of GA, or one-fourth of AO.

Hence, *the centre of gravity of a triangular pyramid is on a line drawn from its vertex to the centre of gravity of its base, and at one-fourth of the distance from the base to the vertex.*

Either face of a triangular pyramid may be taken as the base, the opposite vertex being considered as the vertex of the pyramid.

To find the centre of gravity of a polygonal pyramid; let A-$BCDEF$, represent any pyramid, A being the vertex. Conceive it divided into triangular pyramids, having a common vertex at A. If a plane be passed parallel to the base, and at one-fourth of the distance from the base to the vertex, it follows, from what has just been shown, that the centres of gravity of all the partial pyramids will lie in this plane. We may regard each pyramid as having its weight concentrated at its centre of gravity; hence, the centre of gravity of the entire pyramid must lie in this plane (Art. 52). But it may be shown, as in the case of the triangular pyramid, that the centre of gravity lies somewhere in the line drawn from the vertex to the centre of gravity of the base; it must, therefore, lie where this line pierces the auxiliary plane:

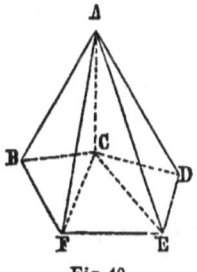

Fig 40.

Hence, *the centre of gravity of any pyramid whatever lies on a line drawn from its vertex to the centre of gravity of its base, and at one-fourth of the distance from the base to the vertex.*

A cone is a pyramid having an infinite number of faces:

Hence, *the centre of gravity of a cone is on a line drawn from the vertex to the centre of gravity of the base, and at one-fourth of the distance from the base to the vertex.*

Centre of Gravity of Prisms and Cylinders.

60. Any prism whatever may be regarded as made up of layers parallel to the bases. If a straight line be drawn between the centres of gravity of the two bases, it will pass through the centres of gravity of all these layers. The centre of gravity of the prism will, therefore, lie somewhere in this line, which we may call the axis of the prism. We may also regard the prism as made up of material lines parallel to the lateral edges of the prism. If a plane be passed midway between the two bases and parallel to them, it will bisect all of these lines, and consequently their centres of gravity, as well as that of the entire prism, will lie in it. It must, therefore, be at the point in which the plane cuts the axis of the prism, that is, at its middle point.

Hence, *the centre of gravity of a prism is at the middle point of its axis.*

When the bases of the prism become polygons having an infinite number of sides, the prism will become a cylinder, and the principle just demonstrated will still hold good:

Hence, *the centre of gravity of a cylinder with parallel bases is at the middle point of its axis.*

Centre of Gravity of Polyhedrons.

61. If any point within a polyhedron be assumed, and this point be joined with each vertex of the polyhedron, we shall thus form as many pyramids as the solid has faces: the centres of gravity of these pyramids may be found by the rules for such cases. If the centres of gravity of the first and second pyramid be joined by a straight line, the common centre of gravity of the two may be found by a process entirely similar to that used in finding the centre of gravity of a polygon, observing that the weights of the partial pyramids are proportional to their volumes, and that they may be represented by their volumes. Having compounded the weights of the first and second, and found its point of application, we may, in like manner, compound this with the weight of the third, and so on, till the centre of gravity of the entire pyramid is determined.

Any solid body bounded by a curved surface may be regarded as a polyhedron whose faces are extremely small, and its centre of gravity may be determined by the rule just explained.

Experimental determination of the Centre of Gravity.

63. We know that the weight of a body always passes through its centre of gravity, no matter what may be the position of the body. If we attach a flexible cord to a body at any point and suspend it freely, it must ultimately come to a state of rest. In this position, the body is acted upon by two forces: the weight, tending to draw the body towards the centre of the earth, and the tension of the cord, which resists this force. In order that the body may be in equilibrium, these forces must be equal and directly opposed. But the direction of the weight passes through the centre of gravity of the body; hence, the tension of the string, which acts in the direction of the string, must also pass through the same point. This principle gives rise to the following method of finding the centre of gravity of a body.

Let ABC represent a body of any form whatever. Attach a string to any point, C, of the body, and suspend it freely; when the body comes to a state of rest, mark the direction of the string; then suspend the body by a second point, B, as before, and when it comes to rest, mark the direction of the string; their point of intersection, G, will be the centre of gravity of the body.

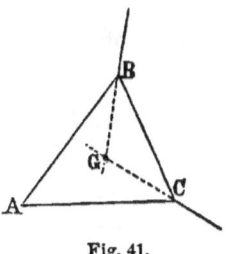

Fig. 41.

Instead of suspending the body by a string, it may be balanced on a point. In this case, the weight acts vertically downwards, and is resisted by the reaction of the point; hence, the centre of gravity must lie vertically over the point.

If, therefore, the body be balanced at any two points of its surface, and verticals be drawn through the point, in these positions, their intersection will be the centre of gravity of the body.

It follows, from what has just been explained, that when a body is suspended by an axis, it can only come to a state of rest when the centre of gravity lies in a vertical plane passed through the axis.

The centre of gravity may lie above the axis, below the axis, or on the axis.

In the first case, if the body be slightly deranged, it will continue to revolve till the centre of gravity falls below the axis; in the second case, it will return to its primitive position; in the third case, it will remain in the position in which it is placed. These cases will be again referred to, under the head of *Stability*.

The preceding rules enable us to find the centres of gravity of all lines, surfaces, and solids; but, on account of the difficulty of applying them in certain cases, we shall annex an outline of some of the methods, by the Differential and Integral Calculus. Those magnitudes whose centres of gravity are most readily found by the calculus, are mathematical curves; areas bounded wholly, or in part, by these curves; curved surfaces; and volumes bound by curved surfaces.

Determination of the Centre of Gravity by means of the Calculus.

64. To place Formulas (13) under a suitable form for the application of the calculus, we have simply to substitute for the forces P, P', &c., the elementary volumes, or the differentials of the magnitudes, and to replace the sign of summation, Σ, by that of integration, $\int$.

Making these changes, Formulas (13) become,

$$x_1 = \frac{\int x\,dm}{\int dm}, \quad y_1 = \frac{\int y\,dm}{\int dm}, \quad z_1 = \frac{\int z\,dm}{\int dm} \cdot \quad (32.)$$

In which dm denotes the differential of the magnitude in question; x, y, and z, the co-ordinates of its centre of gravity, and x_1, y_1, and z_1, the co-ordinates of the centre of gravity of the magnitude.

Application to plane curves.

65. The plane XY may be taken to coincide with that of the curve, in which case, $z = 0$ for every point of the curve; and, consequently, $z_1 = 0$; dm becomes the differential of an arc of a plane curve, or $dm = \sqrt{dx^2 + dy^2}$. Substituting in (32), we have,

$$x_1 = \frac{\int x\sqrt{dx^2 + dy^2}}{\int \sqrt{dx^2 + dy^2}}, \quad y_1 = \frac{\int y\sqrt{dx^2 + dy^2}}{\int \sqrt{dx^2 + dy^2}}. \quad (33.)$$

Centre of Gravity of an arc of a circle.

66. Let ABC be the arc, O the origin of co-ordinates and centre of the circle, OX the axis of abscissas, perpendicular to the chord of the arc, and OY the axis of ordinates. Since the arc is symmetrically situated with respect to the axis of X, the centre of gravity is somewhere on this line (Art. 54); consequently, $y_1 = 0$. To find x_1, we have the equation of the circle,

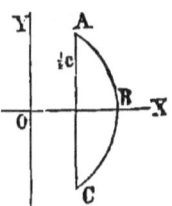

Fig. 42.

$$x^2 + y^2 = r^2.$$

Differentiating,

$$2xdx + 2ydy = 0; \quad \therefore \quad dx^2 = \frac{y^2}{x^2}dy^2.$$

Substituting in the first of Formulas (33), and reducing, we find,

$$x_1 = \frac{\int rdy}{\int \dfrac{rdy}{\sqrt{r^2 - y^2}}}.$$

Integrating both numerator and denominator between the limits $y = -\frac{1}{2}c$, and $y = +\frac{1}{2}c$, we have,

$$\int_{-\frac{1}{2}c}^{+\frac{1}{2}c} rdy = rc;$$

4

and,

$$\int_{-\frac{1}{2}c}^{+\frac{1}{2}c} \frac{r dy}{\sqrt{r^2 - y^2}} = r \sin^{-1} \frac{c}{2r} - r \sin^{-1} \frac{-c}{2r} = \text{arc } ABC.$$

Hence, by substitution,

$$x_1 = \frac{rc}{\text{arc } ABC}; \text{ or, arc } ABC : c :: r : x_1.$$

That is, *the centre of gravity of an arc of a circle is on the diameter which bisects its chord, and its distance from the centre is a fourth proportional to the arc, chord, and radius.*

Application to Plane areas.

67. Let the plane XY be taken to coincide with that of the area. We shall have, as before, $z_1 = 0$. In this case, we have $dm = y dx$; and, consequently, Formulas (32), reduce to

$$x_1 = \frac{\int xy dx}{\int y dx}, \text{ and } y_1 = \frac{\int y^2 dx}{\int y dx} \quad . \quad (34.)$$

Centre of Gravity of a parabolic area.

68. Let AOB represent the area, having its chord at right angles to the axis. Let O be the origin of co-ordinates, taken at the vertex, and let the axis of X coincide with the axis of the curve; the value of y_1 will, as before, be equal to 0. To find the value of x_1, we have the equation of the parabola,

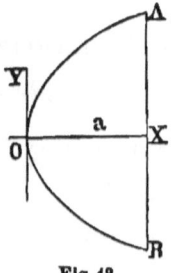

Fig. 43.

$$y^2 = 2px \quad \therefore \quad y = \sqrt{2p} \cdot x^{\frac{1}{2}}.$$

By substitution in the first of Formulas (34), we have,

$$x_1 = \frac{\int \sqrt{2p} \cdot x^{\frac{3}{2}} dx}{\int \sqrt{2p} \cdot x^{\frac{1}{2}} dx} = \frac{\int x^{\frac{3}{2}} dx}{\int x^{\frac{1}{2}} dx}.$$

Integrating between the limits $x = 0$, and $x = a$, we have,

$$\int_0^a x^{\frac{3}{2}} dx = \tfrac{2}{5} a^{\frac{5}{2}},$$

and,

$$\int_0^a x^{\frac{1}{2}} dx = \tfrac{2}{3} a^{\frac{3}{2}}.$$

hence,

$$x_1 = \tfrac{3}{5} a.$$

That is, *the centre of gravity of a segment of a parabola is on its axis, and at a distance from the vertex equal to three-fifths of the altitude of the segment.*

Application to solids of revolution.

69. If we take the planes XY and XZ passing through the axis of revolution, the centre of gravity will lie in both these planes, therefore y_1 and z_1 will both be 0. In this case, the first of Formulas (32) will be sufficient.

Since the axis of X coincides with the axis of revolution, dm becomes equal to $\pi y^2 dx$. Substituting in the first of Formulas (32), we have,

$$x_1 = \frac{\int xy^2 dx}{\int y^2 dx} \quad \ldots \ldots \quad (35.)$$

Centre of Gravity of a semi-ellipsoid.

70. Let the semi-ellipse ACB, be revolved about the axis OC; it will generate a semi-ellipsoid whose axis coincides with the axis of X. Both y_1 and z_1 being 0, it only remains to find the value of x_1.

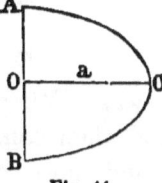

Fig. 44.

The equation of the ellipse referred to its centre, is,

$$y^2 = \frac{b^2}{a^2}(a^2 - x^2),$$

in which a and b are the semi-axes.

Substituting, in Equation (35), we have,

$$x_1 = \frac{\int \frac{b^2}{a^2}(a^2 x - x^3)dx}{\int \frac{b^2}{a^2}(a^2 - x^2)dx} = \frac{\int (a^2 x - x^3)dx}{\int (a^2 - x^2)dx}.$$

Integrating between the limits, $x = 0$, and $x = a$, we have

$$\int_0^a (a^2 x - x^3)dx = \left(\frac{a^4}{2} - \frac{a^4}{4}\right) = \frac{a^4}{4};$$

and,

$$\int_0^a (a^2 - x^2)dx = \left(a^3 - \frac{a^3}{3}\right) = \frac{2}{3}a^3.$$

Substituting, we have,

$$x_1 = \frac{a^4}{4} \div \frac{2a^3}{3} = \frac{3}{8}a = \frac{3}{16} \times 2a.$$

That is, *the centre of gravity of a semi-prolate spheroid of revolution is on its axis of revolution, and at a distance from the centre equal to three-sixteenths of the major axis of the generating ellipse.*

The examples above given are enough to indicate the method of applying the calculus to the determination of the centre of gravity.

Centre of Gravity of a system of bodies.

71. When we have several bodies, and it is required to find their common centre of gravity, it will, in general, be found most convenient to employ the principle of moments.

CENTRE OF GRAVITY. 77

To do this, we first find the centre of gravity of each body separately, by the rules already given. The weight of each body may then be regarded as a force applied at the centre of gravity of the body. The weights being parallel, we have a system of parallel forces, whose points of application are known. If these points are all in the same plane, we may find the lever arms of the resultant of all the weights, with respect to two lines, at right angles to each other in that plane; and these will make known the point of application of the resultant, or, what is the same thing, the centre of gravity of the system. If the points are not in the same plane, the lever arms of the resultant of all the weights may be found, with respect to three axes, at right angles to each other; these will make known the point of application of the resultant weight, or the required position of the centre of gravity.

MISCELLANEOUS EXAMPLES.

1. Required the point of application of the resultant of three equal weights, applied at the three vertices of a plane triangle.

SOLUTION.

Let ABC (Fig. 34) represent the triangle. The resultant of the weights applied at B and C will be applied at D, the middle point of BC. The weight acting at D being double that at A, the total resultant will be applied at G, making $GA = 2GD$; hence, *the required point is at the centre of gravity of the triangle.*

2. Required the point of application of the resultant of a system of equal parallel forces, applied at the vertices of any regular polygon?

Ans. At the centre of gravity of the polygon.

3. Parallel forces of 3, 4, 5, and 6 lbs., are applied at the successive vertices of a square, whose side is 12 inches. At what distance from the first vertex is the point of application of their resultant?

MECHANICS.

SOLUTION.

Take the sides of the square through the first vertex as axes of moments; call the side through the first and second vertex the axis of X, and that through the first and fourth the axis of Y. We shall have from Formulas (13),

$$y_1 = \frac{4 \times 12 + 5 \times 12}{18} = 6;$$

and

$$x_1 = \frac{6 \times 12 + 5 \times 12}{18} = \frac{22}{3},$$

Denoting the required distance by d, we have,

$$d = \sqrt{x_1^2 + y_1^2} = 9.475 \text{ in. } Ans.$$

4. Seven equal forces are applied at seven of the vertices of a cube. What is the distance of the point of application of their resultant from the eighth vertex?

SOLUTION.

Take the eighth vertex as the origin of co-ordinates, and the three edges passing through it as axes of moments. We shall have from Equations (13), denoting one edge of the cube by a,

$$x_1 = \tfrac{4}{7}a, \quad y_1 = \tfrac{4}{7}a, \quad \text{and } z_1 = \tfrac{4}{7}a.$$

Denoting the required distance by d, we have,

$$d = \sqrt{x_1^2 + y_1^2 + z_1^2} = \tfrac{4}{7}a \sqrt{3}. \quad Ans.$$

5. Two isosceles triangles are constructed on opposite sides of the base b, having altitudes respectively equal to h and h', h being greater than h'. Where is the centre of gravity of the space lying within the two triangles?

SOLUTION.

It must lie on the altitude of the greater triangle. Take the common base as an axis of moments; then will the moments of the triangles be, respectively, $\tfrac{1}{2}bh \times \tfrac{1}{3}h$, and

$\frac{1}{2}bh' \times \frac{1}{3}h'$; and from the first of Formulas (13), we shall have,

$$x_1 = \frac{\frac{1}{6}b(h^2 - h'^2)}{\frac{1}{2}b(h + h')} = \frac{1}{3}(h - h').$$

That is, the required centre of gravity is on the altitude of the greater triangle, at a distance from the common base equal to one-third of the difference of the two altitudes.

6. When is the centre of gravity of the space included between two circles tangent to each other internally?

SOLUTION.

Take their common tangent as an axis of moments. The centre of gravity will lie on the common normal, and its distance from the point of contact is given by the equation,

$$x_1 = \frac{\pi r^3 - \pi r'^3}{\pi r^2 - \pi r'^2} = \frac{r^2 + rr' + r'^2}{r + r'}.$$

7. Let there be a square, and suppose it divided by its diagonals into four equal parts, one of which is removed. Required the distance of the centre of gravity of the remaining figure from the opposite side of the square.

Ans. $\frac{7}{18}$ of the side of the square.

8. To construct a triangle, having given its base and centre of gravity.

SOLUTION.

Draw through the middle of the base, and the centre of gravity, a straight line; lay off beyond the centre of gravity a distance equal to twice the distance from the middle of the base to the centre of gravity. The point thus found is the vertex.

9. Given the base and altitude of a triangle. Required the triangle, when its centre of gravity is perpendicularly over the extremity of the base.

10. Three men carry a cylindrical bar, one taking hold

of one end, and the others at a common point. Required the position of this point, in order that the three may sustain equal portions of the weight.

Pressure of one body upon another.

72. Let A be a movable body pressed against a fixed body B, and touching it at a single point. In order that A may be in equilibrium, the resultant of all the forces acting upon it, including its weight, must pass through the point of contact, P'; otherwise there would be a tendency to rotation about P', which would be measured by the moment of the resultant with respect to this point. Furthermore, the direction of the resultant must be normal to the surface of B at the point P', else the body A would have a tendency to slide along the body B, which tendency would be measured by the tangential component. The pressure upon B develops a latent force of reaction, which must be equal and directly opposed to it. The resultant of all the forces must be equal to zero (Art. 47). That is, *when a body, resting upon another and acted upon by any number of forces is in equilibrium, the resultant of all the forces called into play is equal to* 0.

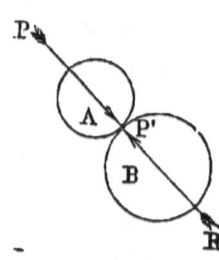

Fig. 45.

If all the forces called into play are taken into account, *the algebraic sums of their moments with respect to any three rectangular axes will be separately equal to* 0.

Equations (29) and (30) are, then, perfectly general in every case of equilibrium, provided all of the forces called into play are taken into account.

Stable, Unstable, and Indifferent Equilibrium.

73. A body is in *stable* equilibrium when, on being slightly disturbed from its state of rest it has a tendency to

STABILTY.

return to that state. This will, in general, be the case when the centre of gravity of the body is at its lowest point. Let A be a spherical body suspended from an axis O, about which it is free to turn. When the centre of gravity of A lies vertically below the axis, it is in equilibrium, for the weight of the body is exactly counterbalanced by the resistance of the axis. Moreover, the equilibrium is *stable;* for if the body be deflected to A', its weight tends to restore it to its position of rest, A. The measure of this tendency is $W \times OP$, that is, the moment of the weight with respect to the axis O. Under the action of the force W, the body will return to A, and, passing to the other side by virtue of its inertia, will finally come to rest and return again to A', and so on, till after a few vibrations, when it will come to rest at A.

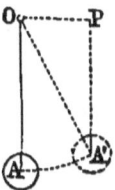

Fig. 46.

A body is in *unstable* equilibrium when, being slightly disturbed from its state of rest, it tends to depart still farther from it. This will, in general, be the case when the centre of gravity of the body occupies its highest position.

Let A be a sphere, connected by an inflexible rod with the axis O. When the centre of gravity of A lies vertically above O, it will be in *unstable* equilibrium; for, if the sphere be deflected to the position A', its weight will act with the lever arm OP to increase this deflection. The motion will continue till, after a few vibrations, it comes to rest below the axis. In this last position, it will be in stable equilibrium.

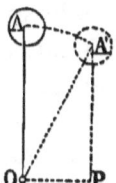

Fig. 47.

A body is in *indifferent*, or *neutral*, equilibrium when it remains at rest wherever it may be placed. This will, in general, be the case when the centre of gravity continues in the same horizontal plane on being slightly disturbed.

4*

Let A be a sphere, supported by a horizontal axis OP passing through its centre of gravity. Then, in whatever position it may be placed, it will have no tendency to change this position; it is, therefore, in *indifferent*, or *neutral* equilibrium.

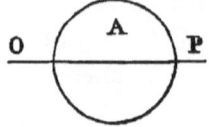

Fig. 48.

In the figure, A, B, and C represent a cone in positions of *stable*, *unstable*, and *indifferent* equilibrium.

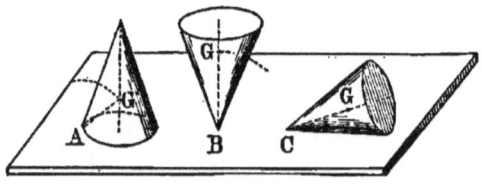

Fig. 49.

If a wheel, or other solid, be mounted on a horizontal axis, about which it is free to turn, the centre of gravity not lying on the axis, it will be in stable equilibrium, when the centre of gravity is directly below the axis; and in unstable equilibrium when it is directly above the axis. When the axis passes through the centre of gravity, it will, in every position, be in neutral equilibrium.

We infer, then, from the preceding discussion, that when a body at rest is so situated that it cannot be disturbed from its position without raising its centre of gravity, it is in a state of *stable equilibrium;* when a slight disturbance depresses the centre of gravity, it is in a state of *unstable equilibrium;* when the centre of gravity remains constantly in the same horizontal plane, it is in a state of *neutral equilibrium*.

This principle holds true in combinations of wheels, as in machinery, and indicates the importance of balancing the elements, so that their centres of gravity may remain as nearly as possible in the same horizontal planes.

Stability of Bodies on Horizontal Planes.

74. A body resting on a horizontal plane may touch it in one, or in more than one point. In the latter case, the salient polygon, formed by joining the extreme points of contact, as *abcd*, is called the *polygon of support*.

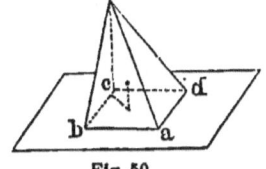

Fig. 50.

When the direction of the weight of the body, that is, the vertical through its centre of gravity, pierces the plane within the polygon of support, the body is *stable*, and will remain in equilibrium, unless acted upon by some other force than the weight of the body. In this case, the body will be most easily overturned about that side of the polygon of support which is nearest to the line of direction of the weight. The moment of the weight, with respect to this side, is called the *moment of stability* of the body. Denoting the weight of the body by W, the distance from the line of direction of the weight to the nearest side of the polygon of support, by r, and the moment of stability by S, we have,

$$S = Wr.$$

The moment of stability is equal to the least moment of any extraneous force which is capable of overturning the body in any direction. The weight of the body remaining the same, its *stability* will increase with r. If the polygon of support is a regular polygon, the stability will be greatest, other things being equal, when the direction of the weight passes through its centre. The area of the polygon of support remaining constant, the stability will be greater as the polygon approaches a circle. The polygon of support being regular, but variable in area, the stability will increase as this area increases. Hence, low bodies resting on extended bases, are, other things being equal, more stable than high bodies resting on narrow bases.

When the direction of the weight passes without the polygon of support, the body is *unstable*, and unless sup ported by some other force than the weight, it will overturn about that side which is nearest to the direction of the weight. In this case, the product of the weight into the shortest distance from its direction to any side of the polygon, is called *the moment of instability*. Denoting this moment by I, we have, as before,

$$I = Wr.$$

'The moment of instability is equal to the least moment of any force which can be applied to prevent the body from overturning.

If the direction of the weight intersect any side of the polygon of support, the body will be *in a state of equilibrium bordering on rotation about that side.*

The stability of a body will be greater, the more nearly the resultant of all the forces acting upon it, including its weight, is to being normal to the bearing surface. A maximum stability will be obtained, other things being equal, when the resultant is exactly perpendicular to the bearing surface. These principles find application in most of the arts, but more especially in Engineering and Architecture. In structures of all kinds intended to be stable, the foundation should be as broad as is consistent with the general design of the work, that the polygon of support may be as great as possible. The pieces for transmitting pressures should be so combined that the pressures transmitted to the ultimate polygons of support should be as nearly normal to the bearing surfaces as possible, and their lines of direction should pass as near the centres of the polygons of support as may be. The same principles hold good at all the points of junction between pieces employed for transmitting pressures. Hence, joints should be made as nearly normal to the pressures as possible.

In the construction of machinery the preceding principles also apply. The centres of gravity of the rotating pieces should be on their axes, otherwise there will result an irregularity of motion, which, besides making the machine

STABILITY. 85

work imperfectly, will ultimately destroy the parts of the machine itself.

In loading cars, wagons, &c., we should endeavor to throw the centre of gravity of the load as near the track as possible. This is, in practice, partially effected by placing the heavier articles at the bottom of the load.

It is needless to enumerate the multitudinous applications of the principles of stability; they are of continual occurrence in the daily transactions of life.

PRACTICAL PROBLEMS IN CONSTRUCTION.

1. A horizontal beam AB, which sustains a load, is supported upon a pivot at A, and by a cord DE, the point E being vertically over A. Required the tension of the cord DE, and the vertical pressure on the pivot A.

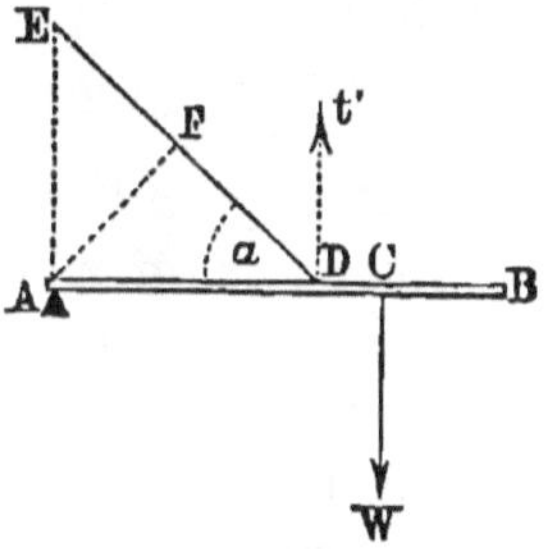

Fig. 51.

SOLUTION.

Denote the weight of the beam, together with its load, by W, and suppose its point of application to be at C. Denote CA by p, and the perpendicular distance AF, from A to DE, by p'. Denote also the tension of the cord by t. If we regard A as the centre of moments, we shall have, in the case of an equilibrium,

$$Wp = tp'; \quad \therefore \quad t = W\frac{p}{p'}.$$

Or, denoting the angles EDA by α, and the distance AD by b, we shall have,

$$t = W\frac{p}{b \sin \alpha}.$$

To find the vertical pressure on the pivot A, resolve the force t into two components, respectively parallel and per pendicular to AB. We shall have for the latter component, denoted by t',

$$t' = t\sin\alpha = W\frac{p}{b}.$$

The vertical pressure upon A, plus the weight W, must be equal to this value of t'. Denoting this pressure by P, we shall have,

$$P + W = W\frac{p}{b}; \text{ or, } P = W\left(\frac{p}{b} - 1\right) = W\left(\frac{p-b}{b}\right);$$

or, $$P = W\frac{DC}{AD}.$$

When $DC = 0$; or, when D and C coincide, the vertical pressure becomes 0.

2. A rope AD, supports a pole, DO, of uniform thickness, one end of which rests upon a horizontal plane, and from the other end is suspended a weight W. Required the tension of the rope, and the thrust, or pressure, on the pole, the weight of the pole being neglected.

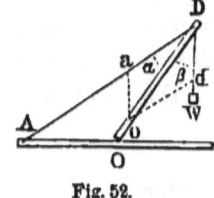

Fig. 52.

SOLUTION.

Denote the tension of the rope by t, the pressure on the pole by p, the angle ADO by α, and the angle ODW by β.

There are three forces acting at D, which hold each other in equilibrium; the weight W, acting downwards, the tension of the rope acting from D, towards A, and the thrust of the pole acting from O towards D. Lay off Dd, to represent the weight, and complete the parallelogram of forces $doaD$; then will Da represent the tension of the rope, and Do the thrust on the pole.

From Art. 35, we have,

$$t \quad W :: \sin\beta : \sin\alpha ; \quad \therefore \quad t = W\frac{\sin\beta}{\sin\alpha}.$$

We have, also, from the same principle,

$$p : W :: \sin(\alpha + \beta) : \sin\alpha ; \quad \therefore \; p = W\frac{\sin(\alpha + \beta)}{\sin\alpha}.$$

If the rope is horizontal, we shall have $\alpha = 90° - \beta$, which gives,

$$t = W\tan\beta, \text{ and } p = \frac{W}{\cos\beta}.$$

3. A beam AB, is suspended by two ropes attached at its extremities, and fastened to pins A and H. Required the tensions upon the ropes.

SOLUTION.

Denote the weight of the beam and its load by W, and suppose that C is the point of application of this force. Denote the tension of the rope BH, by t, and that of the rope FA, by t'. The forces acting to produce an equilibrium, are W, t, and t'. The plane of these forces must be vertical, and further, the directions of the forces must intersect in a point. Produce AF, and BH, till they intersect in K, and draw KC; lay off KC, to represent the weight of the beam and its load, and complete the parallelogram of forces, $KbCf$; then will Kb represent t, and Kf will represent t'. Denote the angle CKB by α, and the angle CKF by β. We shall have, as in the last problem,

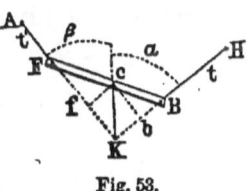

Fig. 53.

$$W : t :: \sin(\alpha + \beta) : \sin\beta ; \quad \therefore \; t = W\frac{\sin(\alpha + \beta)}{\sin\beta}.$$

And,

$$W : t' :: \sin(\alpha + \beta) : \sin\alpha ; \quad \therefore \; t' = W\frac{\sin(\alpha + \beta)}{\sin\alpha}.$$

4. A gate AH, is supported at O upon a pivot, and at A by a hinge, attached to a post AB. Required the pressure on the pivot, and also the tension of the hinge.

SOLUTION.

Denote the weight of the gate and its load, by W. Produce the vertical through the point of application C, of the force W, till it intersects the horizontal through A in D, and draw the line DO. Then will DA and DO represent the directions of the required components of W. Lay off Dc, to represent the value of W, and complete the parallelogram of forces, $Dcoa$; then will Dc represent the pressure on the pivot O, and Da the pressure on the hinge, A. Denoting the angle oDc by α, the pressure on the pivot by p, and on the hinge by p', we shall have,

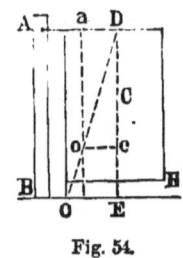

Fig. 54.

$$p = \frac{W}{\cos\alpha}, \text{ and } p' = \frac{p}{\sin\alpha}.$$

If we denote the distance OE by b, and the distance DE by h, we shall have,

$$\cos\alpha = \frac{h}{\sqrt{b^2 + h^2}}, \text{ and } \sin\alpha = \frac{b}{\sqrt{b^2 + h^2}}.$$

Hence,

$$p = \frac{W\sqrt{b^2 + h^2}}{h}, \text{ and } p' = \frac{p\sqrt{b^2 + h^2}}{b}.$$

5. Having given the two rafters AC and BC of a roof, abutting in notches of a tie-beam AB, it is required to find the pressure, or *thrust*, upon the rafters, and the direction and intensity of the pressure upon the joints at the tie-beam.

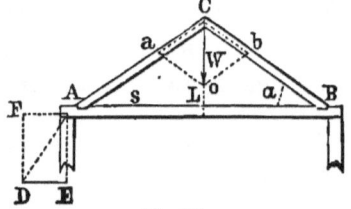

Fig. 55.

SOLUTION.

Denote the weight of the rafters and their load by $2w$; we may regard this weight as made up of three parts—a

weight w, applied at C, and two equal weights $\frac{1}{2}w$, applied at A and B respectively. Let us denote the half span AL by s, the rise CL by h, and the length of the rafter AC or CB by l. Denote, also, the pitch of the roof CBL by α, the thrust on the rafter by t, and the resultant pressure at each of the joints A and B by p.

Lay off Co to represent the weight w, and complete the parallelogram of forces $Cboa$; then will Ca and Cb represent the thrust upon the rafters; and, since the figure $Cboa$ is a rhombus, we shall have,

$$t \sin\alpha = \tfrac{1}{2}w \quad \therefore \quad t = \frac{w}{2\sin\alpha} = \frac{wl}{2h}.$$

Conceive the force t to be applied at A, and resolve it into two components respectively parallel to CL and LA; we shall have for these components,

$$t \sin\alpha = \tfrac{1}{2}w, \quad \text{and} \quad t \cos\alpha = \frac{ws}{2h}.$$

The latter component gives the strain on the tie-beam, AB.

To find the pressure on the joint, we have, acting downwards, the forces $\frac{1}{2}w$ and $\frac{1}{2}w$, or the single force w, and, acting from L towards A, the force $\dfrac{ws}{2h}$; hence,

$$p = \sqrt{w^2 + \frac{w^2 s^2}{4h^2}} = \frac{w}{2h}\sqrt{4h^2 + s^2}.$$

If we denote the angle DAE by β, we shall have from the right-angled triangle DAE,

$$\tan\beta = \frac{DE}{AE} = \frac{s}{2h}.$$

The direction of the joint should be perpendicular to that of the force p, that is, it should make with the horizon an angle whose tangent equals $\dfrac{s}{2h}$.

6. In the last problem suppose the rafters to abut against the wall. Required the least thickness that must be given to the wall to prevent it from being overturned.

SOLUTION.

Denote the entire weight thrown upon the wall by w, the length of that portion of the wall which sustains the pressure p by l', its height by h', its thickness by x, and the weight of each cubic foot of the material of the wall by w'; then will the weight of this part of the wall be equal to $w'h'l'x$.

The force $\dfrac{ws}{2h}$ acts with an arm of lever h' to overturn the wall about its lower and outer edge; this force is resisted by the weight $w + w'h'l'x$, acting through the centre of gravity of the wall with a lever arm equal to $\tfrac{1}{2}x$. If there is an equilibrium, the moments of these two forces must be equal,

that is, $\dfrac{ws}{2h} \times h' = (w + w'h'l'x)\dfrac{x}{2}$, or $\dfrac{wsh'}{h} = wx + w'h'l'x^2$.

Reducing, we have, $x^2 + \dfrac{w}{w'h'l'}x = \dfrac{ws}{w'l'h}$;

or, $x = -\dfrac{w}{2w'h'l'} \pm \sqrt{\dfrac{ws}{w'hl'} + \dfrac{w^2}{4w'^2h'^2l'^2}}$.

7. A sustaining wall has a cross section in the form of a trapezoid, the face upon which the pressure is thrown being vertical, and the opposite face having a slope of *six* perpendicular to *one* horizontal. Required the least thickness that must be given to the wall at the top, that it may not be overturned by a horizontal pressure, whose point of application is at a distance from the bottom of the wall equal to one-third of its height.

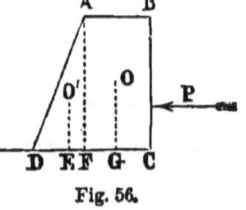

Fig. 56.

SOLUTION.

Pass a plane through the edge A parallel to the face BC, and consider a portion of the wall whose length is one

STABILITY. 91

foot. Denote the pressure upon this portion by P, the height of the wall by $6h$, its thickness at the top by x, and the weight of a cubic foot of the material by w. Let fall from the centres of gravity O and O' of the two portions, the perpendiculars OG and $O'E$, and take the edge D as an axis of moments. The weight of the portion $ABCF$ is equal to $6whx$, and its lever arm, DG, is equal to $h + \tfrac{1}{2}x$. The weight of the portion ADF is $3wh^2$, and its lever arm, DE, is $\tfrac{2}{3}h$. In case of an equilibrium, the sum of the moments of their weights must be equal to the moment of P, whose lever arm is $2h$. Hence,

$$6whx(h + \tfrac{1}{2}x) + 3wh^2 \times \tfrac{2}{3}h = P \times 2h;$$

or, $\qquad 6whx + 3wx^2 + 2wh^3 = 2P.$

Whence, $\qquad x^2 + 2hx = \dfrac{2(P - wh^2)}{3w};$

$$\therefore\ x = -h \pm \sqrt{\dfrac{2(P - wh^2)}{3w} + h^2}.$$

8. Required the conditions of stability of a square pillar acted upon by a force oblique to the axis of the pillar, and applied at the centre of gravity of the pillar's upper base.

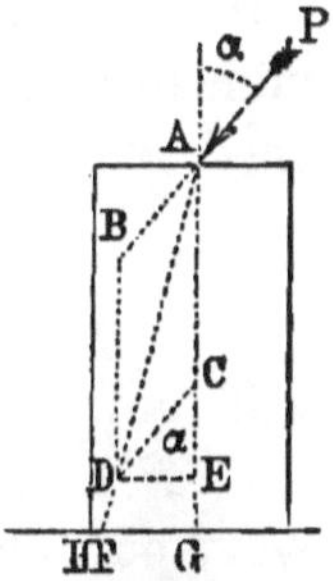

SOLUTION.

Denote the intensity of the oblique force by P, its inclination to the vertical by α, the length or breadth of the pillar by $2a$, its height by x, and the weight of the pillar by W. Through the centre of gravity of the pillar draw the vertical AC, and lay off AC equal to W; prolong PA and lay off AB equal to P; complete the parallelogram of forces $ABDC$, and prolong the diagonal till it intersects HG or HG produced. If the point F falls between H and G, the pillar will be stable; if it falls at H, it will be indifferent; if it falls without H, it will be unstable. To find an expression for the

Fig. 57.

distance FG, draw DE perpendicular to AG. From the similar triangles ADE and AFG, we have,

$$AE : AG :: DE : FG; \quad \therefore FG = \frac{AG \times DE}{AE};$$

But $AG = x$, $DE = P\sin\alpha$, and $AE = W + P\cos\alpha$, hence we have,

$$FG = \frac{Px \sin\alpha}{W + P\cos\alpha};$$

And, since HG equals a, we have the following conditions for stability, indifference and instability, respectively,

$$a > \frac{Px \sin\alpha}{W + P\cos\alpha};$$

$$a = \frac{Px \sin\alpha}{W + P\cos\alpha};$$

$$a < \frac{Px \sin\alpha}{W + P\cos\alpha};$$

If we denote the distance FG by y, and the weight of a cubic foot of the material of the pillar by W, we shall have, since $W = 4a^2xw$,

$$y = \frac{P\sin\alpha \, x}{4a^2wx + P\cos\alpha}.$$

If, now, we suppose the intensity and direction of the force P to remain the same, whilst x is made to assume every possible value from 0 up to any assumed limit, the value of y will undergo corresponding changes. The successive points thus determined make up a line which is called the *line of resistance*, and whose equation is that just deduced.

If the pillar is made up of uncemented blocks, it will remain in equilibrium so long as each joint is pierced by the line of resistance, provided that the tangent to the line of resistance makes with the normal to the joint an angle less than the limiting angle of resistance (Art. 103).

STABILITY.

The highest degree of stability will be attained when the line of resistance is normal to every joint, and when it passes through the centre of gravity of each.

9. To determine the conditions of equilibrium and stability of an arch of uncemented stones.

SOLUTION.

Let $MNLK$ represent half of an arch sustained in equilibrium by a horizontal force P, and by the weight of the archstones. Through the centre of gravity of the first arch-stone draw a vertical line, and on it lay off a distance to represent the weight of that stone. Prolong the direction of P, and lay off a distance equal to the horizontal pressure.

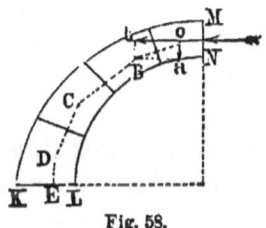

Fig. 58.

Complete the parallelogram of forces, $aobB$, and draw the diagonal oB. This will be the resultant of the forces combined. Combine this resultant with the weight of the second arch-stone, and this with the weight of the third, and so on, till the last inclusive. The polygon $oBCDE$, thus found, is the line of resistance, and if this lies wholly within the solid part of the arch, the arch will be stable; but, if it does not lie within it, the arch will be unstable. A rupture will take place at the joint where the line of resistance passes without the solid part of the arch.

This problem may be solved analytically, in accordance with the principles already illustrated. It is only intended to indicate the general method of proceeding.

CHAPTER IV.

ELEMENTARY MACHINES.

Definitions and General Principles.

75. A MACHINE is a contrivance by means of which a force applied at one point is made to produce an effect at some other point.

The force applied is called the *power*, and the point at which it is applied, is called *the point of application*. The force to be overcome is called *the resistance*, and the point at which it is to be overcome is called *the working point*.

The working of any machine requires a continued application of power. The source of this power is called the *motor*.

Motors are exceedingly various. Some of the most important are *muscular effort*, as exhibited by man and beast in various kinds of work; the *weight and living force of water*, as exhibited in the various kinds of water-mills; the *expansive force of vapors and gases*, as displayed in steam and caloric engines; the *force of air in motion*, as exhibited in the windmill, and in the propulsion of sailing vessels; the *force of magnetic attraction and repulsion*, as shown in the magnetic telegraph and various magnetic machines; the *elastic force of springs*, as shown in watches and various other machines. Of these motors, the most important ones are *steam, air,* and *water power*.

To work, is to exert a certain pressure through a certain distance. The measure of the quantity of work performed by any force, is the product obtained by multiplying the effective pressure exerted, by the distance through which it is exerted.

Machines serve simply to transmit and modify the action of forces. They add nothing to the work of the motor; on the contrary, they absorb and render inefficient much of the work that is impressed upon them. For example, in the case of a water-mill, only a small portion of the *work* expended by the motor is transmitted to the machine, on account of the imperfect manner of applying it, and of this portion a very large fraction is absorbed and rendered practically useless by the various resistances, so that, in reality, only a small fractional portion of the work expended by the motor becomes *effective*.

Of the *applied work*, a part is expended in overcoming *friction, stiffness of cords, bands,* or *chains, resistance of the air, adhesion of the parts,* &c. This goes to *wear out* the machine. A second portion is expended in overcoming sudden impulses, or shocks, arising from the nature of the work to be accomplished, as well as from the imperfect connection of the parts, and from the want of hardness and elasticity in the connecting pieces. This also goes to *strain and wear out* the machine, and also to increase the sources of waste already mentioned. There is often a waste of work arising from a greater supply of motive power than is required to attain the desired result. Thus, in the movement of a train of cars on a railroad, the excess of the work of the steam, above what is just necessary to bring the train to the station, is wasted, and has to be consumed by the application of brakes, an operation which not only wears out the brakes, but also, by creating shocks, injures and ultimately destroys the cars themselves.

Such are some of the sources of the loss of work. A part of these may, by judicious combinations and appliances, be greatly diminished; but, under the most favorable circumstances, there must be a continued loss of work, which requires a continued supply of power from the motor.

In any machine, the quotient obtained by dividing the quantity of *useful*, or *effective work*, by the quantity of *applied work*, is called the *modulus* of the machine. As the resistances are diminished, the modulus increases, and the machine becomes more perfect. Could the modulus ever become equal to 1, the machine would be absolutely *perfect*. Once set in motion, it would continue to move forever, realizing the solution of the problem of *perpetual motion*. It is needless to state that, until the laws of nature are changed, no such realization need be looked for.

In studying the principles of machines, we proceed by approximation. For a first result, it is usual to neglect the effect of hurtful resistances, such as friction, adhesion, stiffness of cords, &c. Having found the relations between the power and resistance under this hypothesis, these relations are afterwards modified, so as take into account the various resistances. We shall, therefore, in the first instance, regard cords as destitute of weight and thickness, perfectly flexible, and inextensible. We shall also regard bars and connecting pieces as destitute of weight and inertia, and perfectly rigid; that is, incapable of compression or extension by the forces to which they may be subjected.

Elementary Machines.

76. The elementary machines are seven in number— viz., the *cord;* the *lever;* the *inclined plane;* the *pulley*, a combination of the cord and lever; the *wheel* and *axle*, also a combination of the cord and lever; the *screw*, a combination of two inclined planes twisted about an axis; and the *wedge*, a simple combination of two inclined planes. It may easily be seen that there are in reality but three elementary machines—the cord, the lever, and the inclined plane. It is, however, more convenient to consider the seven above-named as *elementary*. By a suitable combination of these seven elements, the most complicated pieces of mechanism are produced.

The Cord.

77. Let AB represent a cord solicited by two forces P and R, applied at its extremities, A and B. In order that the cord may be in equilibrium, it is evident, in the first place,

Fig. 59.

that two forces must act in the direction of the cord, and in such a manner as to stretch it, otherwise the cord would bend under the action of the forces. In the second place, the intensities of the forces must be equal, otherwise the greater force would prevail, and motion would ensue. Hence, in order that two forces applied at the extremities of a cord may be in equilibrium, *the forces must be equal and directly opposed.*

The measure of the *tension* of the cord, or *the force by which any two of its adjacent particles are urged to separate,* is the intensity of one of the equal forces, for it is evident that the middle point of the cord might be fixed and either force withdrawn, without diminishing or increasing the tension. When a cord is solicited in opposite directions by unequal forces directed along the cord, the tension will be measured by the intensity of the lesser force.

Let AB represent a cord solicited by two groups of forces applied at its two extremities. In order that these forces may be in equilibrium, the resultant of the group applied at A and the resultant of

Fig. 60.

the group at B must be *equal and directly opposed.* Hence, if we suppose all of the forces at each point to be resolved into components respectively coinciding with, and at right angles to AB, *the normal components at each of the points must be such as to maintain each other in equilibrium, and the resultants of the remaining components at each of the points A and B must be equal and directly opposed.*

Let $ABCD$ represent a cord, at the different points A, B, C, D, of which are applied groups of forces. If these forces are in equilibrium through the intervention of the cord, there must necessarily be an equilibrium at each point of application. Denote the tension of AB, BC, CD, by t, t', t'',

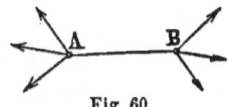

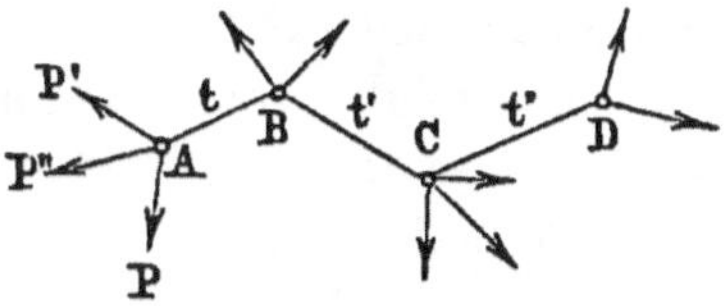

Fig. 61.

and the forces applied by P, P', P'', &c., as shown in the figure. The forces in equilibrium about the point A are P, P', P'', and t, directed from A to B; the forces in equilibrium about B are P''', P^{iv}, t, directed from B to A, and t', directed from B to C. The tension t is the same at all points of the branch AB, and, since it acts at A in the direction AB, and at B in the direction BA, it follows that these two forces exactly counterbalance each other. If, therefore, the forces P, P', P'', were transferred from A to B, unchanged in direction and intensity, the equilibrium at that point would be undisturbed. In like manner, it may be shown that, if all the forces now applied at B be transferred to C, without change of direction or intensity, the equilibrium at C would be undisturbed, and so on to the last point of the cord. Hence we conclude, *that a system of forces applied in any manner at different points of a cord will be in equilibrium, when, if applied at a single point without change of intensity or direction, they will maintain each other in equilibrium.*

Hence, we see that cords in machinery simply serve to transmit the action of forces, without in any other manner modifying their effects.

The Lever.

78. A *lever* is an inflexible bar, free to turn about an axis. This axis is called the *fulcrum*.

Levers are divided into three classes, according to the relative positions of the points of application of the power and resistance.

In the *first* class, the resistance is beyond both the power and fulcrum, and on the side of the fulcrum. The common weighing-scale is an example of this class of levers. The matter to be weighed is the resistance, the counterpoising weight is the power, and the axis of suspension is the fulcrum.

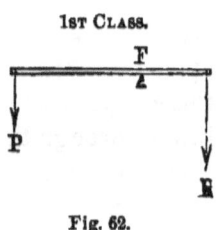

Fig. 62.

In the *second* class, the resistance is between the power and the fulcrum. The oar used in rowing a boat is an example of this class of levers. The end of the oar in the water is the fulcrum, the point at which the oar is fastened to the boat is the point of application of the resistance, and the remaining end of the oar is the point of application of the power.

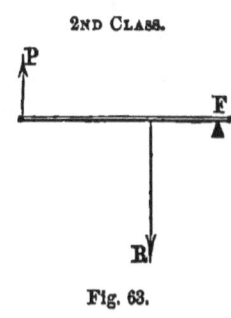

Fig. 63.

In the *third* class, the resistance is beyond both the fulcrum and the power, and on the side of the power. The treadle of a lathe is an example of a lever of this kind. The point at which it is fastened to the floor is the fulcrum, the point at which the foot is applied is the point of application of the power, and the point where it is attached to the crank is the point of application of the resistance.

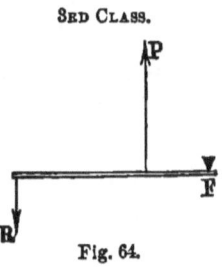

Fig. 64.

Levers may be either *curved* or *straight*, and the directions of the power and resistance may be either parallel or oblique to each other. We shall suppose the power and resistance to be situated in planes at right angles to the fulcrum; for, if they were not so situated, we might conceive each to be resolved into two components—one at right angles, and the other parallel to the axis. The latter component would be exerted to bend the lever laterally, or to make it slide along the axis, developing only hurtful resistance, whilst the former only would tend to turn the lever about the fulcrum.

The perpendicular distances from the fulcrum to the lines of direction of the power and resistance, are called the *lever arms* of these forces. In the bent lever MFN, the perpendicular distances FA and FB are, respectively, the lever arms of P and R.

To determine the conditions of equilibrium of the lever, let us denote the power by P, the resistance by R, and their respective lever arms by p and r. We have the case of a body restrained by an axis, and if we take this as the axis of moments, we shall have for the condition of equilibrium (Art. 49),

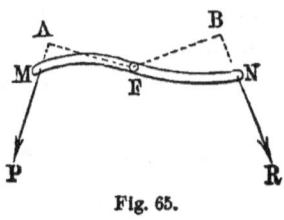

Fig. 65.

$$Pp = Rr\,;\ \text{or,}\ P : R :: r : p\ .\ .\ (36.)$$

That is, *the power is to the resistance, as the lever arm of the resistance is to the lever arm of the power.*

This relation holds good for every kind of lever.

The ratio of the power to the resistance when in equilibrium, either statical or dynamical, is called the *leverage*, or *mechanical advantage*.

When the power is less than the resistance, there is said to be *a gain of power, but a loss of velocity;* that is, the space passed over by the power in performing any work, is as many times greater than that passed over by the resistance, as the resistance is greater than the power. When the power is greater than the resistance, there is said to be *a loss of power, but a gain of velocity.* When the power and resistance are equal, there is neither gain nor loss of power, but simply a change of direction.

In levers of the first class, there may be either a gain or a loss of power; in those of the second class, there is always a gain of power; in those of the third class, there is always a loss of power. A gain of power is always attended with a corresponding loss of velocity, and the reverse.

If several forces act upon a lever at different points, all being perpendicular to the direction of the fulcrum, they will be in equilibrium, when *the algebraic sum of their moments, with respect to the fulcrum, is equal to* 0.

This principle enables us to take into account the weight of the lever, which may be regarded as a vertical force applied at the centre of gravity.

The pressure on the fulcrum is equal to the resultant of the power and resistance, together with the weight of the lever, when that is considered, and it may be found by the rule for finding the resultant of forces applied at points of a rigid body.

The Compound Lever.

79. A compound lever consists of a combination of simple levers AB, BC, CD, so arranged that the resistance in one acts as a power in the next, throughout the combination. Thus, a power P produces at B a resistance R', which, in turn, produces at C a resistance R'', and so on. Let us assume the notation of the figure.

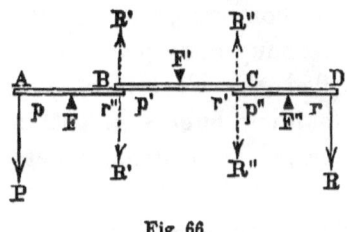

Fig. 66.

From the principle of the simple lever, we shall have the relations,

$$Pp = R'r'', \quad R'p' = R''r', \quad R''p'' = Rr.$$

Multiplying these equations together, member by member, and striking out the common factors, we have,

$$Ppp'p'' = Rrr'r''; \text{ or, } P : R :: rr'r'' : pp'p''. \quad (37.)$$

We might proceed in a similar manner, were there any number of levers in the combination.

Hence, in the compound lever, *the power is to the resistance as the continued product of the alternate arms of lever, commencing at the resistance, is to the continued product of the alternate arms of lever, commencing at the power.*

By suitably adjusting the simple levers, any amount of mechanical advantage may be obtained.

The following combination is used where a great pressure is to be exerted through a very small distance:

The Elbow-joint Press.

80. Let CA, BD, and DE represent bars, with hinge joints at B and D. The bar CA, has a fulcrum at C, and the bar DE works through a guide between D and E. When A is depressed, DE is forced against the upright F, so as to compress, with great force, any body placed between E and F. This machine is called the *elbow-joint press*, and is used in printing, in moulding bullets, in striking coins and medals, in punching holes, riveting steam boilers, &c.

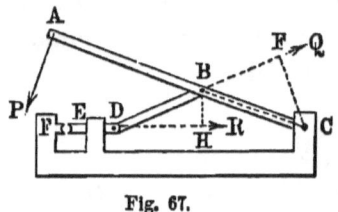

Fig. 67.

Let P denote the force applied at A, perpendicular to AC, Q the resistance in the direction DB, and R the component of Q, in the direction ED. Let C be taken as an axis of moments, and then, because P and Q are in equilibrium, we shall have,

$$P \times AC = Q \times FC, \text{ or, } Q = P \times \frac{AC}{FC}.$$

If we draw BH perpendicular to DR, we shall have,

$$\cos BDH = \frac{DH}{DB};$$

but we have, for the component R,

$$R = Q \cos BDH = Q \times \frac{DH}{DB}.$$

Substituting for Q its value, and reducing,

$$\frac{R}{P} = \frac{AC}{FC} \times \frac{DH}{DB}.$$

When B is depressed, DH and DB approach equality, and FC continually diminishes; that is, the mechanical advantage increases, and finally, when B reaches ER, it becomes infinite. There is no limit to the pressure exerted at F, except that fixed by the strength of the machine.

The Balance.

81. A Balance is a machine for weighing bodies: it consists of a lever AB, called the beam, a knife edge fulcrum F, and two scale-pans D and E, suspended by knife-edges from the extremities of the lever arms FB and FA. These arms should be symmetrical, and of equal length; the knife-edges A, B, and F, should all lie in the same plane, and be perpendicular to a plane through their middle points and the centre of gravity of the beam; they are, therefore, parallel to each other. This condition of parallelism in the same plane, is of essential importance.

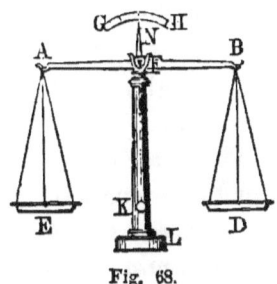

Fig. 68.

In addition to this, the middle points of the knife-edges A, B, and F, should be on the same straight line, perpendicular to the plane through the fulcrum F, and the centre of gravity of the beam. The knife-edges should be of hardened steel, and their supports should either be of polished agate, or, what is still better, of hardened steel, so as to diminish the effect of friction along the lines of contact. The fulcrum may be made horizontal, by leveling-screws passing through the foot-plate L. A needle N, projects upwards, or sometimes downwards, which, playing in front of a graduated arc GH, serves to show the deflection of the line of knife-edges from the horizontal. When the instrument is not in use, the fulcrum may be raised from its bearings by a pinion K, working into a rack in the interior of the standard FK. The knife-edges A and B may, by a similar arrangement, be raised from their bearings also.

The ordinary balances of the shops are similar in their general plan; but many of the preceding arrangements are omitted. The scale-pans being exactly alike, the balance should remain in equilibrium, with the line AB horizontal, not only when the balance is without a *load*, but also when the pans are loaded with equal weights; and when AB is deflected from the horizontal, it should return to this position. This result is attained by throwing the centre of gravity slightly below the line AB. To test a balance, let two weights be placed in the pans that will exactly counterbalance each other, then change the weights to the opposite pans; if the equilibrium is still maintained, the balance is said to be *true*.

The *sensibility* of a balance is its capability of indicating small differences of weight. The sensibility will be greater, *as the lengths of the arms increase, as the centre of gravity of the beam approaches the fulcrum, as the mass of the load decreases, and as the length of the needle increases.* The centre of gravity of the beam being below the fulcrum, it may be made to approach to or recede from it, by a solid ball of metal attached to the beam by means of a screw, by which it may be raised or depressed at pleasure. The remaining conditions of sensibility will be limited by the strength of the material, and the use to which it is to be applied.

Should it be found that a balance is not true, it may still be employed, with but slight error, as indicated below.

Denote the length of the lever arms, by r and r', and the weight of the body, by W. When the weight W is applied at the extremity of the arm r, denote the counterpoising weights employed, by W'; and when it is applied at the extremity of the arm r', denote the counterpoising weights employed, by W''. We shall have, from the principle of the lever,

$$Wr = W'r', \text{ and } Wr' = W''r.$$

Multiplying these equations, member by member, we have,

$$W^2 rr' = W''W'rr'; \quad \therefore \quad W = \sqrt{W'W''};$$

that is, *the true weight is equal to the square root of the product of the apparent weights.*

A still better method, and one that is more free from the effects of errors in construction, is to place the body to be weighed in one scale and add counterpoising weights till the beam is horizontal; then remove the body to be weighed and replace it by known weights till the beam is again horizontal; the sum of the replacing weights will be the weight required. If, in changing the loads, the positions of the knife-edges are not moved, this method is almost exact, but this is a condition difficult to fulfill in manipulation.

The Steelyard.

82. The steelyard is an instrument used for weighing bodies. It consists of a lever AB, called the beam; a fulcrum F; a scale-pan D, attached at the extremity of one arm; and a known weight E, movable along the other arm. We shall suppose the weight of E to be 1 lb. This instrument is sometimes more convenient than the balance, but it is more inaccurate. The conditions of sensibility are essentially the same as for the balance. To graduate the instrument, place a pound-weight in the pan D, and move the counterpoise E till the beam rests horizontal—let that point be marked 1; next place a 10 lb. weight in the pan, and move the counterpoise E till the beam is again horizontal, and let that point be marked 10; divide the intermediate space into nine equal parts, and mark the points of division as shown in the figure. These spaces may be subdivided at pleasure, and the scale extended to any desirable limits. We have supposed that the centre of gravity coincides with the fulcrum; when this is not the case, the weight of the instrument must be taken into account as a force applied at its centre of gravity. We may then graduate the beam by experiment, or we may compute the lever arms, corresponding to the different weights, by the general principle of moments.

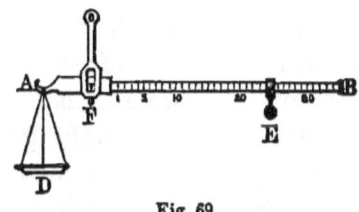

Fig. 69.

To weigh any body with the steelyard, place it in the scale-pan and move the counterpoise E along the beam till an equilibrium is established between the two; the corresponding mark on the beam will indicate the weight.

The bent Lever Balance.

83. This balance consists of a bent lever ACB; a fulcrum C; a scale-pan D; and a graduated arc EF, whose centre coincides with the centre of motion C. When a weight is placed in the scale-pan, the pan is depressed and the lever-arm of the weight is diminished; the weight B is raised, and its lever-arm increased. When the moments of the two forces become equal, the instrument will come to a state of rest, and the weight will be indicated by a needle projecting from B, and playing in front of the arc FE. The zero of the arc EF is at the point indicated by the needle when there is no load in the pan D.

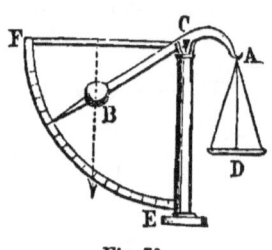

Fig. 70.

The instrument may be graduated experimentally by placing weights of 1, 2, 3, &c., pounds in the pan, and marking the points at which the needle comes to rest, or it may be graduated by means of the general principle of moments. We need not explain this method of graduation.

To weigh a body with the bent lever balance, place it in the scale-pan, and note the point at which the needle comes to rest; the reading will make known the weight sought.

Compound Balances.

84. Compound balances are much used in weighing heavy articles, as merchandise, coal, freight for shipping, &c. A great variety of combinations have been employed, one of which is annexed.

AB is a platform, on which the object to be weighed is

placed; *BC* is a guard firmly attached to the platform; the platform is supported upon the knife-edge fulcrum *E*, and the piece *D*, through the medium of a brace *CD*; *GF* is a lever turning about the fulcrum *F*,

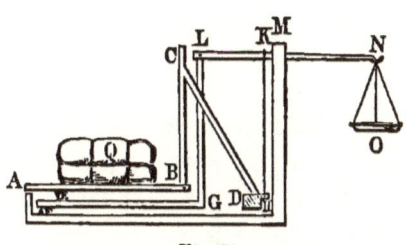

Fig. 71.

and suspended by a rod from the point *L*; *LN* is a lever having its fulcrum at *M*, and sustaining the piece *D* by a rod *KH*; *O* is a scale-pan suspended from the end *N* of the lever *LN*. The instrument is so constructed, that

$$EF : GF :: KM : LM;$$

and the distance *KM* is generally made equal to $\frac{1}{10}$ of *MN*. The parts are so arranged that the beam *LN* shall rest horizontally in equilibrium when no weight is placed on the platform.

If, now, a body *Q* be placed upon the platform, a part of its weight will be thrown upon the piece *D*, and, acting downwards, will produce an equal pressure at *K*. The remaining part will be thrown upon *E*, and, acting upon the lever *FG*, will produce a downward pressure at *G*, which will be transmitted to *L*; but, on account of the relation given by the above proportion, the effect of this pressure upon the lever *LN* will be the same as though the pressure thrown upon *E* had been applied directly at *K*. The final effect is, therefore, the same as though the weight of *Q* had been applied at *K*, and, to counterbalance it, a weight equal to $\frac{1}{10}$ of *Q* must be placed in the scale-pan *O*.

To weigh a body, then, by means of this scale, place it on the platform, and add weights to the scale-pan till the lever *LN* is horizontal, then 10 times the sum of the weight added will be equal to the weight required. By making other combinations of levers, or by combining the principle of the steelyard with this balance, objects may be weighed by using a constant counterpoise.

EXAMPLES.

1. In a lever of the first class, the lever arm of the resistance is $2\frac{2}{3}$ inches, that of the power, $33\frac{1}{3}$, and the resistance 100 lbs. What is the power necessary to hold the resistance in equilibrium? *Ans.* 8 lbs.

2. Four weights of 1, 3, 5, and 7 lbs. respectively, are suspended from points of a straight lever, eight inches apart. How far from the point of application of the first weight must the fulcrum be situated, that the weights may be in equilibrium?

SOLUTION.

Let x denote the required distance. Then, from Art. (36)

$$1 \times x + 3(x - 8) + 5(x - 16) + 7(x - 24) = 0;$$

$$\therefore x = 17 \text{ in. } Ans.$$

3. A lever, of uniform thickness, and 12 feet long, is kept horizontal by a weight of 100 lbs. applied at one extremity, and a force P applied at the other extremity, so as to make an angle of 30° with the horizon. The fulcrum is 20 inches from the point of application of the weight, and the weight of the lever is 10 lbs. What is the value of P, and what is the pressure upon the fulcrum?

SOLUTION.

The lever arm of P is equal to 124 in. $\times \sin 30° = 62$ in., and the lever arm of the weight of the lever is 52 in. Hence,

$$20 \times 100 = 10 \times 52 + P \times 62; \quad \therefore P = 24 \text{ lbs. nearly.}$$

We have, also,

$$R = \sqrt{X^2 + Y^2} = \sqrt{(110 + 24 \sin 30°)^2 + (24 \cos 30°)^2}.$$

$$\therefore R = 123.8 \text{ lbs.;}$$

and, $\quad \cos a = \dfrac{X}{R} = \dfrac{20.785}{123.8} = .16789;$

$$\therefore a = 80° \; 20' \; 02''.$$

4. A heavy lever rests on a fulcrum which is 2 feet from one end, 8 feet from the other, and is kept horizontal by a weight of 100 lbs., applied at the first end, and a weight of 18 lbs., applied at the other end. What is the weight of the lever, supposed of uniform thickness throughout?

SOLUTION.

Denote the required weight by x; its arm of lever is 3 feet. We have, from the principle of the lever,

$100 \times 2 = x \times 3 + 18 \times 8; \quad \therefore x = 18\frac{2}{3}$ lbs. *Ans.*

5. Two weights keep a horizontal lever at rest; the pressure on the fulcrum is 10 lbs., the difference of the weights is 4 lbs., and the difference of lever arms is 9 inches. What are the weights, and their lever arms?

Ans. The weights are 7 lbs. and 3 lbs.; their lever arms are $15\frac{3}{4}$ in., and $6\frac{3}{4}$ in.

6. The apparent weight of a body weighed in one pan of a false balance is $5\frac{1}{2}$ lbs., and in the other pan it is $6\frac{6}{11}$ lbs. What is the true weight?

$$W = \sqrt{\tfrac{11}{2} \times \tfrac{72}{11}} = 6 \text{ lbs. } Ans.$$

7. In the preceding example, what is the ratio of the lever arms of the balance?

SOLUTION.

Denote the shorter arm by l, and the longer arm by nl. We shall have, from the principle of moments,

$6l = 5\frac{1}{2} \times nl,$ or, $6nl = 6\frac{6}{11}l; \quad \therefore n = 1\frac{1}{11}.$

That is, the longer arm equals $1\frac{1}{11}$ times the shorter arm.

The Inclined Plane.

85. An inclined plane is a plane inclined to the horizon. In this machine, let the power be a force applied to a body either to prevent motion down the plane, or to produce motion up the plane, and let the resistance be the weight of the body acting vertically downwards. The power may be applied in any direction whatever; but we shall, for simplicity's sake, suppose it to be in a vertical plane, taken perpendicular to the inclined plane.

Let AB represent the inclined plane, O a body resting on it, R the weight of the body, and P the force applied to hold it in equilibrium. In order that these two forces may keep the body at rest, friction being neglected, their resultant must be perpendicular to AB (Art. 72).

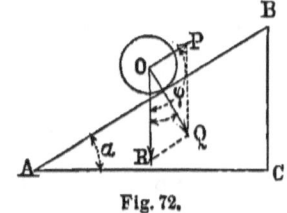

Fig. 72.

When the direction of the force P is given, its intensity may be found geometrically, as follows: draw OR to represent the weight, and OQ perpendicular to AB; through R draw RQ parallel to OP, and through Q draw QP parallel to OR; then will OP represent the required intensity, and OQ the pressure on the plane.

When the intensity of P is given, its direction may be found as follows: draw OR and OQ as before; with R as a centre, and the given intensity as a radius, describe an arc cutting OQ in Q; draw RQ, and through O draw OP parallel, and equal to RQ; it will represent the direction of the force P.

If we denote the angle between P and R by φ, and the inclination of the plane by α, we shall have the angle ROQ equal to α, since OQ is perpendicular to AB, and OR to AC, and, consequently, the angle $QOP = \varphi - \alpha$. From the principle of Art. 35, we have,

$$P : R :: \sin \alpha : \sin(\varphi - \alpha) \quad . \quad . \quad (38.)$$

ELEMENTARY MACHINES.   111

From which, if either P or φ be given, the other can be found.

If we suppose the power to be applied parallel to the plane, we shall have, $\quad \varphi - \alpha = 90°$,
or, $\quad \sin(\varphi - \alpha) = 1$.

We have, also, $\quad \sin \alpha = \dfrac{BC}{AB}$.

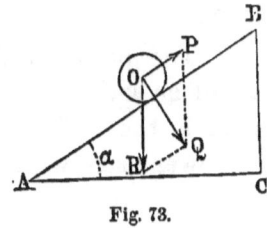

Fig. 73.

Substituting these in the preceding proportion, and reducing, we have,

$$P : R :: BC : AB \quad \ldots \quad (39.)$$

That is, when the power is parallel to the plane, *the power is to the resistance, as the height of the plane is to its length.*

If the power is parallel to the base of the plane, we shall have, $\varphi - \alpha = 90° - \alpha$; whence,

$$\sin(\varphi - \alpha) = \cos \alpha = \dfrac{AC}{AB};$$

also, $\quad \sin \alpha = \dfrac{BC}{AB};$

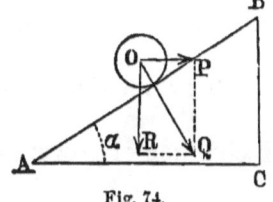

Fig. 74.

Substituting in Proportion (38), and reducing, we have,

$$P : R :: BC : AC \quad \ldots \quad (40.)$$

That is, *the power is to the resistance as the height of the plane is to its base.*

From the last proportion we have,

$$P = R \dfrac{BC}{AC} = R \tan \alpha.$$

If we suppose α to increase, the value of P will increase, and when α becomes $90°$, P will become infinite; that is, if friction be neglected, no finite horizontal force can sustain a body against a vertical wall.

EXAMPLES.

1. A power of 1 lb., acting parallel to an inclined plane, supports a weight of 2 lbs. What is the inclination of the plane?   *Ans.* 30°.

2. The power, resistance, and normal pressure, in the case of an inclined plane, are, respectively, 9, 13, and 6 lbs. What is the inclination of the plane, and what angle does the power make with the plane?

SOLUTION.

If we denote the angle between the power and resistance by φ, and the inclination of the plane by α, we shall have, from Art. (35),

$$6 = \sqrt{13^2 + 9^2 + 2 \times 9 \times 13 \cos \varphi};$$

$$\therefore \varphi = 156° \ 8' \ 20''.$$

Also, from Art. (35), for the inclination of the plane,

$$6 : 9 :: \sin 156° \ 8' \ 20'' : \sin \alpha; \quad \therefore \alpha = 37° \ 21' \ 26''.$$

Inclination of power to plane $= \varphi - 90° - \alpha = 28° \ 46' \ 54''$.
Ans.

3. A body may be supported on an inclined plane by a force of 10 lbs., acting parallel to the plane; but it requires a force of 12 lbs. to support it when the force acts parallel to the base. What is the weight of the body, and what is the inclination of the plane?

Ans. The weight is 18.09 lbs., and the inclination is 33° 33′ 25″.

The Pulley.

86. A pulley consists of a wheel having a groove around its circumference to receive a cord; the wheel turns freely on an axis at right angles to its plane, which axis is supported by a frame called a *block*. The pulley is said to be *fixed*, when the block is fixed, and to be *movable*, when

ELEMENTARY MACHINES.

the block is movable. Pulleys may be used singly, or in combinations.

Single fixed Pulley.

87. In this pulley the block, and, consequently, the axis, is fixed. Denote the power by P, the resistance by R, and the radius of the pulley by r. It is plain that both the power and resistance should be in a plane, at right angles to the axis. Hence, if we take the axis of the pulley as the axis of moments, we shall have (Art. 49), the following condition of equilibrium:

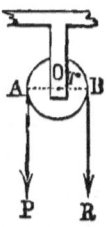

Fig. 75.

$$Pr = Rr; \text{ or, } P = R.$$

That is, in the single fixed pulley, *the power is equal to the resistance.*

The effect of the pulley is, therefore, simply to change the direction of the force, and it is for this purpose that it is generally used.

Single Movable Pulley.

88. In this pulley the block, and, consequently, the axis, is movable. The resistance is applied at a hook attached to the block; one end of a rope, enveloping the lower part of the pulley, is firmly attached at a fixed point C, and the power is applied at the other extremity. We shall take the two branches of the rope parallel, that being the most advantageous way of using the machine.

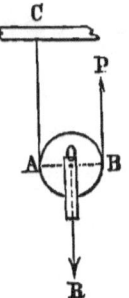

Fig. 76.

Adopting the notation of the preceding article, and taking A, the point of contact of CA with the pulley, as the centre of moments, we shall have, for the condition of equilibrium (Art. 49),

$$P \times 2r = Rr; \quad \therefore \quad P = \tfrac{1}{2}R.$$

That is, in the movable pulley, when the power and resistance are parallel, *the power is equal to one half of the resistance.* The tension upon the cord CA is evidently the same as that upon the cord BP. It is, therefore, equal to the power, or to one-half the resistance. If, therefore, the resistance of the fixed point C be replaced by a force equal to P, the equilibrium will be undisturbed.

If the two branches of the enveloping cord are oblique to each other, the condition of equilibrium will be somewhat modified. Suppose the resistance of the fixed point C to be replaced by a force equal to P, and denote the angle between the two branches of the cord by 2φ. If an equilibrium subsists between the forces P, P, and R, we must have the relation,

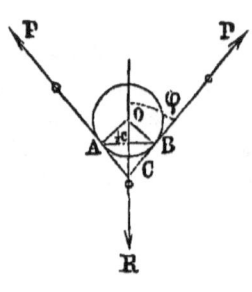

Fig. 77.

$$2P\cos\varphi = R.$$

Draw the chord AB between the points of contact of the cord and pulley, and denote its length by c; draw, also, the radius OB. Then, since OR is perpendicular to AB and BP to OB, the angle ABO will be equal to one half of the angle ACB, or equal to φ. Hence,

$$\cos\varphi = \tfrac{1}{2}c \div r = \frac{c}{2r}.$$

Substituting in the preceding equation and reducing, we have,

$$Pc = Rr\,; \quad \therefore\ P : R\ :: r : c \ \ . \ . \ (41.)$$

That is, *the power is to the resistance as the radius of the pulley is to the chord of the arc enveloped by the rope.*

When the chord is greater than the radius, there will be a gain of *mechanical advantage* in the use of this pulley; when less, there will be a loss of *mechanical advantage.*

If the chord becomes equal to the diameter, we have, as before,

$$P = \tfrac{1}{2}R.$$

Combinations of Separate Movable Pulleys.

89. The figure represents a combination of three movable pulleys, in which there are as many separate cords as there are pulleys; the first end of each cord is attached at a fixed point, the second end being fastened to the hook of the next pulley in order, except the last cord, at the second extremity of which the power is applied.

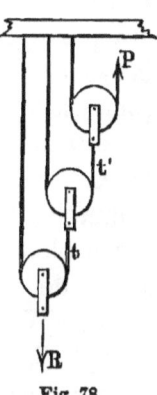

Fig. 78.

Let us denote the tension of the cord between the first and second pulley by t, that of the cord between the second and third pulley by t'. By the preceding Article, we have,

$$t = \tfrac{1}{2}R; \quad t' = \tfrac{1}{2}t; \quad P = \tfrac{1}{2}t'.$$

Multiplying these equations together, member by member, and striking out the common factors in the resulting equation, we have,

$$P = (\tfrac{1}{2})^3 R.$$

Had there been n pulleys in the combination, we should have obtained, in an entirely similar manner, the relation,

$$P = (\tfrac{1}{2})^n . R; \quad \therefore \ P : R :: 1 : 2^n \ . \ (42.)$$

That is, *the power is to the resistance as 1 is to 2^n, n denoting the number of pulleys.*

For convenience, the last branch of the cord is often passed over a fixed pulley; this arrangement only serves to change the direction of the force, without in any way changing the conditions of equilibrium.

Combinations of Pulleys in blocks.

90. These combinations are effected in a variety of ways. In most cases, there is but a single rope employed, which, being firmly attached to a hook of one block, passes around a pulley in the other block, then around one in the first block, and so on, passing from block to block until it has passed around each pulley in the system. The power is applied at the free end of the rope. Sometimes the pulleys in the same block are placed side by side, sometimes they are placed one above another, as represented in the figure, in which case the interior pullies are made somewhat smaller than the outer ones. The conditions of equilibrium are the same in both cases. To deduce the conditions of equilibrium in the case represented, in which the upper block is fixed and the lower one movable: denote the power by P, the resistance by R. When there is an equilibrium between P and R, the tension upon each branch of the rope which aids in supporting the resistance must be the same, and equal to P; but, since the last pulley simply serves to change the direction of the force P, there will be four such branches in the case considered; hence, we shall have,

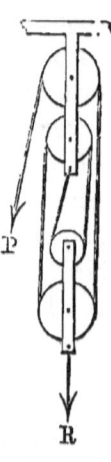

Fig. 79.

$$4P = R, \text{ or } P = \tfrac{1}{4}R.$$

Had there been n pulleys in the combination, there would have been n supporting branches of the cord, and we should have had, in the same manner,

$$nP = R, \text{ or } P : R :: 1 : n \quad . \quad . \quad (43.)$$

That is, *the power is to the resistance as 1 is to the number of branches of the rope which support the resistance.*

The principles involved in the combinations already considered, will be sufficient to make known the relation between the power and resistance in any combination whatever.

EXAMPLES.

1. In a system of six movable pulleys, of the kind described in Art. 89, what weight can be sustained by a power of 12 lbs? *Ans.* 768 lbs.

2. In a combination of pulleys in two blocks, when there are six pulleys in each block, what weight can a power of 12 lbs. sustain in equilibrium ? *Ans.* 144 lbs.

3. In a combination of separate movable pulleys, the resistance is 576 lbs., and the power which keeps it in equilibrium is 9 lbs. How many pulleys are there in the combination ? *Ans.* 6.

4. In a combination of pulleys in two blocks, with a single rope, the power is 62 lbs.; and the resistance 496 lbs. How many pulleys are there in each block ? *Ans.* 4.

5. In a combination of two movable pulleys, the inclinations of the ropes at each pulley is 120°. What is the power required to support a weight of 27 lbs. ? *Ans.* 9 lbs.

The Wheel and Axle.

91. The wheel and axle consists of a wheel A, mounted on an axle or arbor B. The power is applied at one extremity of a rope wrapped around the wheel, and the resistance at one extremity of a second rope, wrapped around the axle in a contrary direction. The whole instrument is supported by pivots projecting from the ends of the axle. In deducing the conditions of equilibrium of the power and resistance, we shall suppose them to be situated in planes, at right angles to the axis.

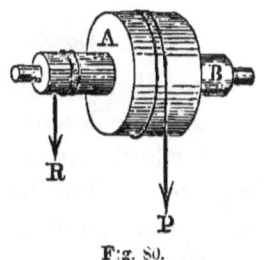

Fig. 80.

Denote the power by P, the resistance by R, the radius of the wheel by r, and the radius of the axle by r'. We shall have, in case of an equilibrium (Art. 49),

$Pr = Rr'$, or $P : R :: r' : r$. (44.)

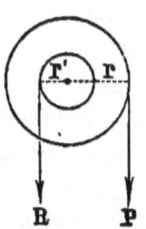

Fig. 81.

That is, *the power is to the resistance as the radius of the axle is to the radius of the wheel.*

By suitably varying the dimensions of the wheel and axle, any amount of *mechanical advantage* may be obtained.

If we draw a straight line from the point of contact of the first rope and the wheel, to the point of contact of the second rope and the axle, the power and resistance being parallel, it can readily be shown that it will cut the axis of revolution at a point which divides the line through the points of contact into two parts, which are inversely proportional to the power and resistance. Hence, this is the point of application of the resultant of these two forces. The resultant will be equal to the sum of the forces, and by the aid of the principle of moments, the pressure on each pivot may be computed. When the weight of the machine is to be taken into account, we must regard it as a vertical force applied at the centre of gravity of the wheel and axle. The pressures upon each pivot due to this weight, may be computed separately, and added to those already found.

Combinations of Wheels and Axles.

92. If the rope of the first axle be passed around a second wheel, and the rope of the second axle around a third wheel, and so on, a combination will result which is capable of affording great mechanical advantage. The figure represents a combination of two wheels and axles. To deduce the conditions of equilibrium, denote the power by P, the resistance by R, the radius of the first wheel by r, that of the first axle by r', that of the second wheel by r'', and that of the second axle by r'''. If we denote the tension of the connecting rope by t, this may be regarded as a power applied to the second wheel. From what was demonstrated for the wheel and axle, we shall have,

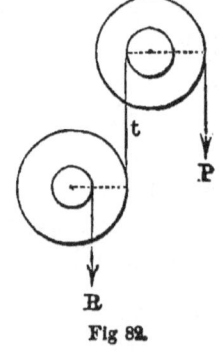

Fig 82.

$$Pr = tr', \text{ and } tr'' = Rr'''.$$

Multiplying these equations together, member by member, and reducing, we have,

$$Prr'' = Rr'r'''; \quad \text{or,} \quad P : R :: r'r''' : rr''.$$

In like manner, were there any number of wheels and axles in the combination, we might deduce the relation,

$$Prr''r^{\text{iv}} \ldots = Rr'r'''r^{\text{v}} \ldots;$$

or, $\quad P : R :: r'r'''r^{\text{v}} \ldots : rr''r^{\text{iv}} \ldots \quad$ (45.)

That is, *the power is to the resistance as the continued product of the radii of the axles is to the continued product of the radii of the wheels.*

The principle just explained, is applicable to those kinds of machinery in which motion is transmitted from wheel to wheel by the aid of bands, or belts. An endless band, called the driving belt, passes around one drum mounted upon the axle of the driving wheel, and around another on that of the driven wheel. When the radius of the former is greater than that of the latter, there is a gain of velocity, and a corresponding loss of power; in the contrary case, there is a loss of velocity, and a corresponding gain of power. In the first case, we are said to *gear up for velocity;* in the second case, we are said to *gear down for power.* These remarks admit of extension to combinations of any number of pieces, in which motion is transmitted by belts, cords, chains, or, as we shall see hereafter, by trains of toothed wheels.

The Crank and Axle, or Windlass.

93. This machine consists of an axle AB, and a crank BCD. The power is applied to the crank-handle DC, and the resistance to a rope wrapped around the axle. The distance from the handle DC to the axis, is called the crank-arm.

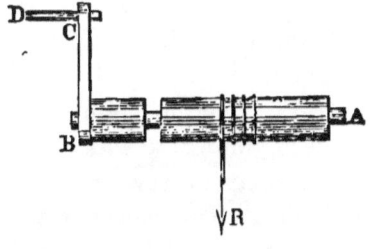

Fig. 83.

The relation between the power and resistance, when in equilibrium, is the same as in the wheel and axle, except that we substitute the crank-arm for the radius of the wheel.

Hence, *the power is to the resistance as the radius of the axle is to the crank-arm.*

This machine is used in drawing water from wells, raising ore from mines, and the like. It is also used in combination with other machines. Instead of the crank, as shown in the figure, two holes are sometimes bored at right angles to each other and to the axis, and levers inserted, at the extremities of which the power is applied. The condition of equilibrium remains unchanged, provided we substitute for the crank-arm, the distance from the point of application of the power to the axis.

The Capstan.

94. The *Capstan* differs in no material respect from the windlass, except in having its axis vertical. The capstan consists of a vertical axle passing through strong guides, and having holes at its upper end for the insertion of levers. It is much used on shipboard for raising anchors. The conditions of equilibrium are the same as in the windlass.

The Differential Windlass.

95. This differs from the common windlass in having an axle formed of two cylinders, A and B, of different diameters, but having a common axis. A rope is attached to the larger cylinder, and wrapped several times around it, after which it passes around the movable pulley C, and, returning, is wrapped in a contrary direction about the smaller cylinder, to which the second end of the rope is made fast. The power is applied at the crank-handle FE, and the resistance to the hook of the movable pulley. When the crank is turned so as to

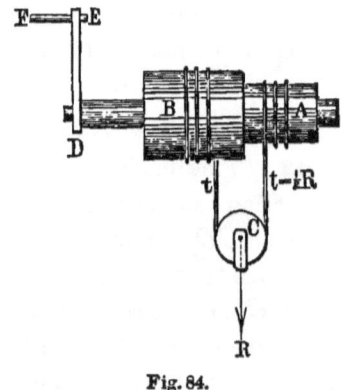

Fig. 84.

wind the rope upon the larger cylinder it unwinds from the smaller one, but in a less degree, and the total effect of the power is to raise the resistance R. To deduce the conditions of equilibrium between the power and resistance, denote the power by P, the resistance by R, the crank-arm by c, the radius of the larger cylinder by r, and that of the smaller cylinder by r'. The resistance acts equally upon the two branches of the rope from which it is suspended, hence the tension of each branch may be represented by $\frac{1}{2}R$. Suppose that the power acts to wind the rope upon the larger cylinder. The moment of the power will be Pc; the moment of the tension of the branch A will be equal to $\frac{1}{2}Rr'$, this acts to assist the power; the moment of the tension of the branch B will be equal to $\frac{1}{2}Rr$, this acts to oppose the power. From the principle of moments, we have,

$$Pc + \tfrac{1}{2}Rr' = \tfrac{1}{2}Rr, \text{ or } Pc = \tfrac{1}{2}R(r - r');$$

whence,

$$P : R :: r - r' : 2c. \quad . \quad . \quad . \quad (46.)$$

That is, *the power is to the resistance as the difference of the radii of the two cylinders is to twice the crank-arm.*

By increasing the crank-arm and diminishing the difference between the radii of the cylinders, any amount of mechanical advantage may be obtained by the use of this machine.

Wheel-work.

96. The principle employed in finding the relation between the power and resistance in a train of wheel-work is the same as that used in discussing the wheel and axle and its modifications. To illustrate the method of proceeding, we have taken the case in which the power is applied to a crank-handle which is attached to the axis of a cogged wheel

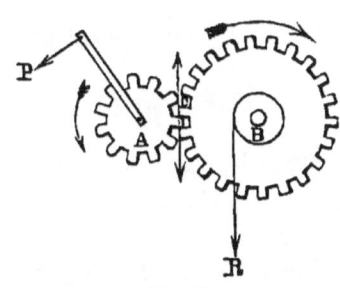

Fig. 85

A; the teeth, or cogs, of this wheel work into the spaces of the toothed wheel B, and the resistance is attached to a rope wound round the arbor of the last wheel. In order that the wheel A may communicate motion freely to the wheel B, the number of teeth in their circumferences should be proportional to their radii, and the spaces between the teeth in one wheel should be large enough to receive the teeth of the other wheel, but not large enough to allow a great deal of play. The teeth should always come in contact at the same distances from the centres of the wheels, and those distances are taken as the radii of the wheels themselves. Denote the power by P, the resistance by R, the crank-arm by c, the radius of the wheel A by r, that of the wheel B by r', that of the arbor by r'', and suppose the power and resistance to be in equilibrium; then will the pressure due to the action of the power tend to turn the wheels in the direction of the arrow heads. This tendency will be counteracted by the pressure of the resistance tending to produce motion in a contrary direction. If we denote the pressure at the point C by R', we should have, from what has preceded,

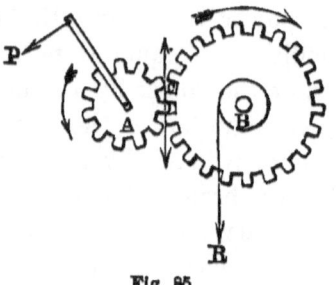

Fig. 85.

$$Pc = R'r \quad \text{and} \quad R'r' = Rr'';$$

whence, by multiplication and reduction,

$$Pcr' = Rrr'', \quad \text{or} \quad P : R :: rr'' : cr' \quad . \quad (47.)$$

That is, *the power is to the resistance as the continued product of the alternate arms of lever, beginning at the resistance, is to the continued product of the alternate arms of lever beginning at the power.*

Had there been any number of wheels in the train lying

between the power and resistance, we should have found similar conditions of equilibrium.

EXAMPLES.

1. A power of 5 lbs., acting at the circumference of a wheel whose radius is 5 feet, supports a resistance of 200 lbs. applied at the circumference of the axle. What is the radius of the axle? *Ans.* $1\frac{1}{2}$ inches.

2. The radius of the axle of a windlass is 3 inches, and the crank-arm 15 inches. What power must be applied to the crank-handle, to support a resistance of 180 lbs., applied to the circumference of the axle? *Ans.* 36 lbs.

3. A power P, acts upon a rope 2 inches in diameter, passing over a wheel whose radius is 3 feet, and supports a resistance of 320 lbs., applied by a rope of the same diameter, passing over an axle whose radius is 4 inches. What is the value of P, when the thickness of the rope is taken into account. *Ans.* $43\frac{9}{37}$ lbs.

The Screw.

97. The screw is essentially a combination of two inclined planes. It consists of a solid cylinder, called *the cylinder of the screw*, which is enveloped by a spiral projection called the *thread*. The thread may be generated as follows: let an isosceles triangle be placed so that its base shall coincide with an element of the cylinder of the screw, and so that its plane shall pass through the axis. Let the triangle be revolved uniformly about the axis, and at the same time be moved uniformly in the direction of the axis, at such a rate that it shall pass over a distance in this direction equal to the base of the triangle during one revolution. The solid generated by the triangle is the thread of the screw. The two sides of the triangle generate helicoidal surfaces, which constitute the upper and lower surfaces of the thread. Every point in these lines generates a curve called a helix, which is entirely similar to an inclined

Fig. 86.

plane bent around a cylinder. The vertex generates what is called the *outer helix*, and the two angular points of the base trace out the same curve, which is the *inner helix*. The screw just described is called a screw with a triangular thread. Had we used a rectangle, instead of a triangle, and imposed the condition, that the motion in the direction of the axis during one revolution, should be equal to twice the base, we should have had a screw with a rectangular thread, as in the figure.

The screw works into a piece called a *nut*, which is generated in a manner entirely analogous to that just described, except that what is solid in the screw is wanting in the nut; it is, therefore, exactly adapted to receive the thread of the screw. Sometimes, the screw remains fast, and the nut is turned upon it; in which case, the nut has a motion of revolution, combined with a longitudinal motion. Sometimes, the nut remains fast, and the screw is turned within it, in which case, the screw receives a motion in the direction of its axis, in connection with a motion of rotation. The conditions of equilibrium are the same for each. In both cases, the power is applied at the extremity of a lever; when in motion, the point of application describes an ascending or descending spiral, resulting from a combination of the rotary and the longitudinal motion. We shall suppose the nut to remain fast, and the screw to be movable, and that the resistance acts parallel to the axis of the screw. If the axis is vertical, and the resistance a weight, we may regard that weight as resting upon one of the helices, and sustained in equilibrium by a force applied horizontally. If we suppose the supporting helix to be developed on a vertical plane, it will form an inclined plane, whose base is the circumference of the base of the cylinder on which it lies, and whose altitude is the distance between the threads of the screw.

Let AB represent the development of this helix on a vertical plane, and denote by F the force applied parallel to the base, and *immediately* to the weight R, to sustain it on the plane. We shall have (Art 85),

$$F : R :: BC : AC.$$

But the power is actually applied through the medium of a lever. Denoting the radius OG of the cylinder of the supporting helix, by r, and the arm of lever of the power P by p, we shall have, from the principle of the lever,

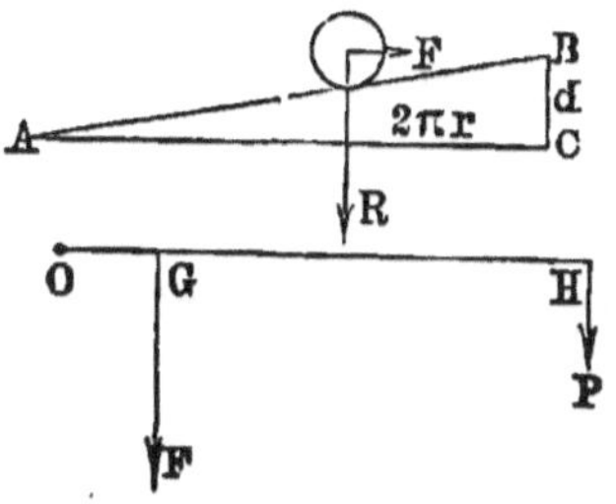

$$P : F :: r : p;$$

Fig. 87.

or, $\qquad P : F :: 2\pi r : 2\pi p.$

Combining this proportion with the preceding one, and recollecting that $AC = 2\pi r$, we deduce the proportion,

$$P : R :: BC : 2\pi p \quad . \quad . \quad . \quad (48.)$$

That is, *the power is to the resistance as the distance between the threads is to the circumference described by the point of application of the power.*

By suitably diminishing the distance between the threads, other things being equal, any amount of *mechanical advantage* may be obtained.

The screw is used for producing great pressures through very small distances, as in pressing books for the binder, packing merchandise, expressing oils, and the like. On account of the great amount of friction, and other hurtful resistances developed, the modulus of the machine is very small.

The Differential Screw.

98. The differential screw consists essentially of an ordinary screw, as just described, into the end of which works a smaller screw, having its axis coincident with the first, but having its thread turned in a contrary direction; that is, it is what is technically called a left-handed screw, the first screw being a right-handed one. The distance between the threads of the second screw is somewhat less than that between the threads of the first screw, and this difference may be made as small as desirable. The second screw is so arranged that it admits of a longitudinal motion, but not of a motion of rotation. By the action of the differential screw, the weight is raised vertically through a distance equal to the difference of the distances between the threads on the two screws, for each revolution of the point of application of the power. For, were the first screw alone to turn, the weight would be raised through a distance equal to the distance between its threads; but, because the second screw is à left-handed one, this distance will be diminished by a distance equal to that between its threads. We may, therefore, write the following rule:

The power is to the resistance as the difference of the distances between the threads of the two screws is to the circumference described by the point of application of the power.

Endless Screw.

99. The endless screw is a screw secured by shoulders, so that it cannot be moved longitudinally, and working into a toothed wheel. The distance between the teeth should be nearly equal to the distance between the threads of the screw. When the screw is turned, it imparts a rotary motion to the wheel, which may be utilized by any mechanical device. The conditions of equilibrium are the same as for the screw, the resistance in this case being offered by the wheel, in the direction of its circumference.

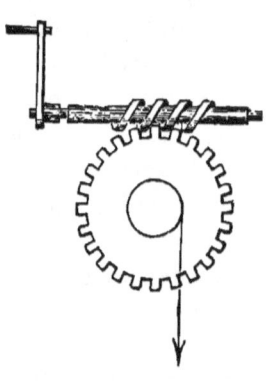

Fig. 88.

Machines of this kind are used in determining the number of revolutions of an axis. An endless screw is arranged to turn as many times as the axis, and being connected with a train of light wheel-work, the last piece of which bears an index, the number of revolutions can readily be ascertained at any instant. As an example, suppose the first wheel to have 100 teeth, and to bear on its arbor a smaller wheel, having 10 teeth; suppose this wheel to engage with a larger wheel having 100 teeth, and so on. When the endless screw has made 10,000 revolutions, the first wheel will have made 100 revolutions, the second large wheel will have made 10 revolutions, and the third wheel 1 revolution. By a suitable arrangement of indices, the exact number of revolutions of the axis, at any instant, may be read off from the instrument.

EXAMPLES.

1. What must be the distance between the threads of a screw in order that a power of 28 lbs., acting at the extremity of a lever 25 inches long, may sustain a weight of 10,000 lbs.? *Ans.* .4396 inches.

2. The distance between the threads of a screw is $\frac{1}{3}$ of an inch. What resistance can be supported by a power of 60 lbs., acting at the extremity of a lever 15 inches long? *Ans.* 16,964 lbs.

3. The distance from the axis of the trunions of a gun weighing 2,016 lbs. to the elevating screw is 3 feet, and the distance of the centre of gravity of the gun from the same axis is four inches. If the distance between the threads of the screw be $\frac{2}{3}$ of an inch, and the length of the lever 5 inches, what power must be applied to sustain the gun in a horizontal position? *Ans.* 4.754 lbs.

The Wedge.

100. The wedge is a solid, bounded by a rectangle BD, called the *back;* two equal rectangles, AF and DF, called *faces;* and two equal isosceles triangles, called *ends.* The line EF, in which the faces meet, is called the *edge.*

The power is applied at the back, to which its direction should be normal, and the resistance is applied to the faces, and in directions normal to them. One half of the resistance

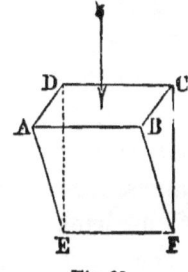

Fig. 89.

is applied normally to one face, and the other half normally to the other face. Let ABC be a section of a wedge made by a plane at right angles to the edge. Denote the power by P, and the resistance opposed to each face by $\frac{1}{2}R$; denote the angle BAC of the wedge by 2φ. Produce the directions of the resistances till they intersect in O. This point will be on the line of direction of the power. Lay off OF to represent the power, and complete

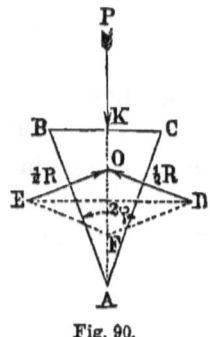

Fig. 90.

the parallelogram ED; then will OE and OD represent the resistances developed by the power. Let each of the forces $\frac{1}{2}R$ be resolved into two components, one perpendicular to OF, and the other coinciding with it. The two former will be equal and directly opposed to each other, whilst the two latter will hold the force P in equilibrium. Since DE is perpendicular to FO, and DO perpendicular to CA, the angle ODE is equal to the angle OAC, or φ. The component of $\frac{1}{2}R$ in the direction of OF, is $\frac{1}{2}R\sin\varphi$; hence, twice this, or $R\sin\varphi = P$. But $\sin\varphi = \dfrac{CK}{CA} = \dfrac{\frac{1}{2}b}{l}$, in which b denotes the breadth of the back BC, and l the length of the face CA. Substituting this expression for $\sin\varphi$, and reducing, we have,

$$R \times \tfrac{1}{2}b = Pl, \text{ or } P : R :: \tfrac{1}{2}b : l \quad . \quad (49.)$$

That is, *the power is to the resistance as one-half of the breadth of the back is to the length of the face of the wedge.*

The mechanical advantage of the wedge may be increased by diminishing the breadth of the back, or, in other words, by making the edge sharper. The principle of the wedge finds an important application in all cutting instruments, as knives, razors, and the like. By diminishing the thickness of the back, the instrument is rendered liable to break, hence the necessity of forming cutting instruments of the hardest and most tenacious materials.

General remarks on Elementary Machines.

101. We have thus far supposed the power and resistance to be in equilibrium, through the intervention of the machine, their points of application being at rest. If we now suppose the point of application to be moved through any distance, by the action of an extraneous force, the point of application of the power will move through a corresponding space. These spaces will be described in conformity with the design of the machine; and it will be found, in each instance, that they are inversely proportional to the forces. If we suppose these spaces to be infinitely small, they may, in all cases, be regarded as straight lines, which will also be the virtual velocities of the forces. If the point of application moves in a direction contrary to the direction of the resistance, the point of application of the power will move in the direction of the power. If we denote the paths described by those points respectively, by δr, and δp, we shall have,

$$P\delta p - R\delta r = 0; \text{ or } P\delta p = R\delta r \quad . . \quad (50.)$$

That is, *the algebraic sum of the virtual moments is equal to 0.* Or, we might enunciate the principle in another manner, by saying, that in all cases, the quantity of work of the power is equal to the quantity of work of the resistance.

We shall illustrate this principle, by considering a single case, that of the single movable pulley, leaving its further application to the remaining machines, as exercises for the student.

In the figure, suppose that an extraneous force acts to raise the resistance R, through the infinitely small space DE, denoted by δr; the point of application of P must be raised through the infinitely small space FG, denoted by δp, in order that the equilibrium may be preserved.

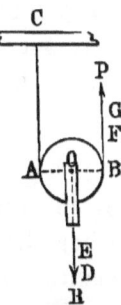

Fig. 91.

In order that the resistance may be raised through the distance DE, both branches of the rope enveloping the pulley must be shortened by the same amount; or, what is the same thing, the free end of the rope must ascend through **twice** the distance DE. Hence,

$$\delta p = 2\delta r.$$

But, from the conditions of equilibrium,

$$P = \tfrac{1}{2}R.$$

Multiplying these equations, member by member, we have,

$$P\delta p = R\delta r.$$

Hence, the principle is proved for this particular case. In like manner, it may be shown to hold good for all of the elementary machines.

The principle of *equality of work* of the power and resistance being true for any infinitely short time, it must necessarily hold good for any time whatever. Hence, we conclude, that the quantity of work of the power, in overcoming any resistance, is equal to quantity of work of the resistance. Although, by the application of a very small power, we are able to overcome a very great resistance, the space passed over by the point of application of the power must be as much greater than that passed over by the point of application of the resistance, as the resistance is greater than the power. This is generally expressed by saying, that *what is gained in power is lost in velocity.*

We see, therefore, that no power is, or can be, gained; the only function of a machine being to enable a smaller force to accomplish in a longer time, what a larger force would be required to perform in a shorter time.

Friction.

102. FRICTION is the resistance which one body experiences in moving upon another, the two being pressed together by some force. This resistance arises from inequalities in the two surfaces, the projections of one surface sinking into the depressions of the other. In order to overcome this resistance, a sufficient force must be applied to break off, or bend down, the projecting points, or else to lift the moving body clear of the inequalities. The force thus applied, is equal, and directly opposed to the force of friction, which is tangential to the two surfaces. The force which presses the surfaces together, is normal to them both at the point of contact.

Friction is distinguished as *sliding* and *rolling*. The former arises when one body is drawn upon another; the latter when one body is rolled upon another. In the case of rolling friction, the motion is such as to lift the projecting points out of the depressions; the resistance is, therefore, much less than in sliding friction.

Between certain bodies, the friction is somewhat different when motion is just beginning, from what it is when motion has been established. The friction developed when a body is passing from a state of rest to a state of motion, is called *friction of quiescence;* that which exists between bodies in motion, is called *friction of motion.*

The following *laws of friction* have been established by numerous experiments, viz.:

First, *the friction of quiescence between the same bodies, is proportional to the normal pressure, and independent of the extent of the surfaces in contact.*

Secondly, *the friction of motion between the same bodies, is proportional to the normal pressure, and independent, both of the extent of the surfaces in contact, and of the velocity of the moving body.*

Thirdly, *for compressible bodies, the friction of quiescence is greater than the friction of motion; for bodies which are sensibly incompressible, the difference is scarcely appreciable.*

Fourthly, *friction may be greatly diminished, by interposing unguents between the rubbing surfaces.*

Unguents serve to fill up the cavities of surfaces, and thus to diminish the resistances arising from their roughness. For slow motions and great pressures, the more consistent unguents are used, as lard, tallow, and various mixtures for rapid motions, and light pressures, oils are generally employed.

The ratio obtained by dividing the entire force of friction by the normal pressure, is called the *coefficient of friction;* the value of the coefficient of friction for any two substances, may be determined experimentally as follows :

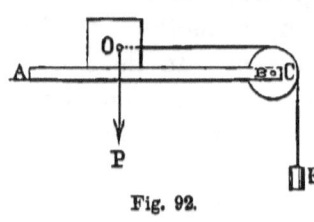

Fig. 92.

Let AB be a horizontal plane formed of one of the substances, and let O be a cubical block of the other substance resting upon it. Attach a string OC, to the block, so that its direction shall pass through its centre of gravity, and be parallel to AB; let the string pass over a fixed pulley C, and let a weight F, be attached to its extremity.

Increase the weight F till the body O just begins to slide along the plane, then will this weight measure the whole force of friction. Denote this weight by F, that of the body, or the normal pressure, by P, and the coefficient of friction, by f. Then, from the definition, we shall have,

$$f = \frac{F}{P}.$$

In this manner, values for f, corresponding to different substances, may be found, and arranged in tables. This experiment gives the friction of quiescence. If the weight F is such as to keep the body O in uniform motion, the resulting value of f will correspond to friction of motion.

The value of f, for any substance, is called the *unit*, or *coefficient* of friction. Hence, we may define the unit, or coefficient of friction, to be *the friction due to a normal pressure of one pound.*

Having given the normal pressure in pounds, and the unit of friction, the entire friction will be found by multiplying these quantities together.

There is a second method of finding the value of f experimentally, as follows:

Let AB be an inclined plane, formed of one of the substances, and O a cubical block, formed of the other substance, and resting upon it. Elevate the plane till the block just begins to slide down the plane by its own weight. Denote the angle of inclination, at this instant, by α, and the weight of O, by W. Resolve the force W into two components, one normal to the surface of the plane, and the other one parallel to it. Denote the former component by P, and the latter by Q. Since OW is perpendicular to AC, and OP to AB, the angle WOP is equal to α. Hence,

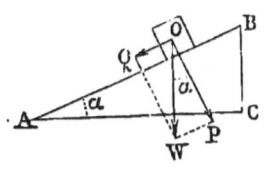

Fig. 93.

$$P = W\cos\alpha, \quad \text{and} \quad Q = W\sin\alpha.$$

The normal pressure being equal to $W\cos\alpha$, and the force of friction being $W\sin\alpha$, we shall have, from the principles already explained,

$$f = \frac{W\sin\alpha}{W\cos\alpha} = \tan\alpha = \frac{BC}{AC}.$$

The angle α is called the angle of friction.

Limiting Angle of Resistance.

103. Let AB be any plane surface, and O a body resting upon it. Let R be the resultant of all the forces acting upon it, including the weight applied at the centre of gravity. Denote the angle between R and the normal to AB, by α, and suppose R to be resolved into two components P and Q, the former parallel to AB, and the latter perpendicular to it; we shall have,

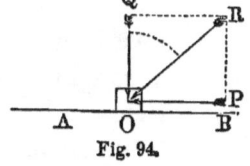

Fig. 94.

$$P = R\sin\alpha, \quad \text{and} \quad Q = R\cos\alpha.$$

The friction due to the normal pressure will be equal to $fR\cos\alpha$. Now, when the tangential component $R\sin\alpha$ is less than $fR\cos\alpha$, the body will remain at rest; when it is greater than $fR\cos\alpha$, the body will slide along the plane; and when the two are equal, the body will be in a state bordering on motion along the plane. Placing the two equal, we have,

$$fR\cos\alpha = R\sin\alpha; \quad \therefore f = \tan\alpha.$$

The value of α is called the limiting angle of resistance, and is equal to the inclination of the plane, when the body is about to slide down by its own weight. If, now, the line OR be revolved about the normal, it will generate a conical surface, within which, if any force whatever, including the weight, be applied at the centre of gravity, the body will remain at rest, and without which, if a sufficient force be applied, the body will slide along the plane. This cone is called the *limiting cone of resistance*.

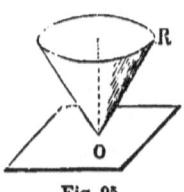

Fig. 95.

The values of f, or the coefficient of friction, in some of the most common cases, as determined by MORIN, is appended:

TABLE.

Bodies between which friction takes place.	Coefficient of friction.
Iron on oak,	.62
Cast iron on oak,	.49
Oak on oak, fibres parallel,	.48
Do., do., greased,	.10
Cast iron on cast iron,	.15
Wrought iron on wrought iron,	.14
Brass on iron,	.16
Brass on brass,	.20
Wrought iron on cast iron,	.19
Cast iron on elm,	.19
Soft limestone on the same,	.64
Hard limestone on the same,	.38

Bodies between which friction takes place	Coefficient of friction
Leather belts on wooden pulleys,	.47
Leather belts on cast iron pulleys,	.28
Cast iron on cast iron, greased,	.10

Pivots or axes of wrought or cast iron, on brass or cast iron pillows:

1st, when constantly supplied with oil,	.05
2nd, when greased from time to time,	.08
3rd, without any application,	.15

Rolling Friction.

104. Rolling friction is the resistance which one body offers to another when rolling along its surface, the two being pressed together by some force. This resistance, like that in sliding friction, arises from the inequalities of the two surfaces. The *coefficient*, or *unit*, of rolling friction is equal to the quotient obtained by dividing the entire force of friction by the normal pressure. This coefficient is much less than the coefficient of sliding friction.

The following laws of friction have been established, when a cylindrical body or wheel rolls upon a plane:

First, the coefficient of rolling friction is proportional to the normal pressure:

Secondly, *it is inversely proportional to the diameter of the cylinder or wheel:*

Thirdly, *it increases as the surface of contact and velocity increase.*

In many cases there is a combination of both sliding and rolling friction in the same machine. Thus, in a car upon a railroad-track, the friction at the axle is sliding, and that between the circumference of the wheel and the track is rolling.

Adhesion.

105. ADHESION is the resistance which one body experiences in moving upon another in consequence of the cohesion existing between the molecules of the surfaces in contact. This resistance increases when the surfaces are allowed to remain for some time in contact, and is very slight when motion has been established. Both theory and experiment show that adhesion between the same surfaces is proportional to the extent of the surface of contact.

The *coefficient of adhesion* is the quotient obtained by dividing the entire adhesion by the area of the surface of contact. Or, denoting the entire adhesion by A, the area of the surface of contact by S, and the *coefficient of adhesion* by a, we have,

$$a = \frac{A}{S}, \quad \text{or} \quad A = aS.$$

To find the entire adhesion, we multiply the unit of adhesion by the area of the surface of contact.

Stiffness of Cords.

106. Let O represent a pulley, with a cord AB, wrapped around its circumference, and suppose a force P, applied at B, to overcome the resistance R, and impart motion to the pulley. As the rope winds upon the pulley, at C, its rigidity acts to increase the arm of lever of R, and to overcome this resistance to flexure an additional force is required. For the same pulley, this additional force may be represented by the algebraic expression,

$$a + bR,$$

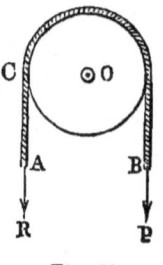

Fig. 96.

in which a and b are constants dependent upon the nature and construction of the rope, and R is the resistance to be overcome, or the tension of the cord AC. The values of a and b for different ropes have been ascertained by experiment, and tabulated. Finally, if the same rope be wound upon pulleys of different diameters, the additional force is found to vary inversely as their diameters. If the diameter of the pulley be denoted by D, and the resistance due to stiffness of cordage be denoted by S, we shall have,

$$S = \frac{a + bR}{D}.$$

In the case of the pulley, if we neglect friction, we shall have, when the motion is uniform,

$$P = R + \frac{a + bR}{D},$$

for the algebraic expression of the conditions of equilibrium.

The values of a and b have been determined experimentally for all values of R and D, and tabulated.

Atmospheric Resistance.

107. The atmosphere exercises a powerful resistance to the motion of bodies passing through it. This resistance is due to the inertia of the particles of air, which must be overcome by the force of a moving body. It is evident, in the first place, other things being equal, that the resistance will depend upon the amount of surface of the moving body which is exposed to the air in the direction of the motion. In the second place, the resistance must increase with the square of the velocity of the moving body; for, if we suppose the velocity to be doubled, there will be twice as many particles met with in a second, and each particle will collide against the moving body with twice the force, hence; if the velocity be doubled, the resistance will be quadrupled. By a similar course of reasoning, it may be shown that, if the velocity be tripled, the retardation will become nine times as great, and so on. If, therefore, the retardation corresponding to a square foot of surface, at any given velocity, be determined, the retardation corresponding to any surface and any velocity whatever may be computed.

Influence of Friction on the Inclined Plane.

108. Let it be required to determine the relation between the power and resistance, when the power is just on the point of imparting motion to a body up an inclined plane, friction being taken into account.

Let AB represent the plane, O the body, OP the power on the point of imparting motion up the plane, and OR the weight of the body. Denote the power by P, the weight by R, the inclination of the plane by α, and the angle between the direction of the power and the normal to the plane by β. Let P and R be resolved into components respectively parallel and perpendicular to the plane.

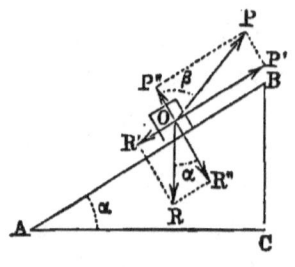

Fig. 97.

We shall have, for the parallel components, $R\sin\alpha$ and $P\sin\beta$, and for the perpendicular components, $R\cos\alpha$ and $P\cos\beta$. The resultant of the normal components will be equal to $R\cos\alpha - P\cos\beta$; and, if we denote the coefficient of friction by f, we shall have for the entire force of friction (Art. 102),

$$f(R\cos\alpha - P\cos\beta).$$

When we consider the body on the eve of motion up the plane, the component $P\sin\beta$ must be equal and directly opposed to the resultant of the force of friction and the component $R\sin\alpha$; hence, we must have,

$$P\sin\beta = R\sin\alpha + f(R\cos\alpha - P\cos\beta).$$

Performing the multiplications indicated, and reducing, we have,

$$P = R\left\{\frac{\sin\alpha + f\cos\alpha}{\sin\beta + f\cos\beta}\right\} \quad . \quad . \quad . \quad (51.)$$

If we suppose an equilibrium to exist, the body being on on the eve of motion down the plane, we shall have.

$$P\sin\beta + f(R\cos\alpha - P\cos\beta) = R\sin\alpha.$$

Whence, by reduction,

$$P = R\left\{\frac{\sin\alpha - f\cos\alpha}{\sin\beta - f\cos\beta}\right\} \quad . \quad . \quad . \quad (52.)$$

From these expressions, two values of P may be found, when α, β, f, and R are given. It is evident that any value of P greater than the first will cause the body to slide up the plane, that any value less than the second will permit it to slide down the plane, and that for any intermediate value the body will remain at rest on the plane.

If we suppose P to be parallel to the plane, we shall have $\sin\beta = 1$, $\cos\beta = 0$, and the two values of P reduce to

$$P = R(\sin\alpha + f\cos\alpha) \quad \ldots \quad (53.)$$

and,

$$P = R(\sin\alpha - f\cos\alpha) \quad \ldots \quad (54.)$$

If friction be neglected, we have $f = 0$; whence, by substitution,

$$P = R\sin\alpha, \quad \text{or} \quad \frac{P}{R} = \frac{BC}{AB};$$

a result which agrees with that deduced in a preceding article.

To find the quantity of work of the power whilst drawing a body up the entire length of the inclined plane, it may be observed that the value of P, in Equation (53), is equal to that required to maintain the body in uniform motion after motion has commenced.

Multiplying both members of that equation by AB, we have,

$$P \times AB = R \times AB\sin\alpha + fR \times AB\cos\alpha$$
$$= R \times BC + fR \times AC.$$

But $R \times BC$ is the quantity of work necessary to raise the body through the vertical height BC; and $fR \times AC$, is the quantity of work necessary to draw the body horizontally through the distance AC (Art. 75). Hence, the quantity of work required to draw a body up an inclined plane, when the power is parallel to the plane, is equal to the quantity of work necessary to draw it horizontally across the base of the plane, *plus* the quantity of work necessary to raise it vertically through the height of the plane.

A curve situated in a vertical plane may be regarded as made up of an infinite number of inclined planes. We infer, therefore, that the quantity of work necessary to draw a body up a curve, the power acting always parallel to the direction of the curve, is equal to the quantity of work necessary to draw the body over the horizontal projection of the curve, plus the quantity of work necessary to raise the body through a height equal to the difference of altitude of the two extremities of the curve.

The last two principles enable us to compare the quantities of work necessary to draw a train of cars over a horizontal track, and up an inclined track, or a succession of inclined tracks. We may, therefore, compute the length of a horizontal track which will consume the same amount of work, furnished by the motor, as is actually consumed in consequence of the undulation of the track.

We are thus enabled to compare the relative advantages of different proposed routes of railroad, with respect to the motive power required for working them.

Line of Least Traction.

109. The force employed to draw a body with uniform motion along an inclined plane, is called the *force of traction;* and the line of direction of this force is the line of traction. In Equation (51), P represents the force of traction required to keep a body in uniform motion up an inclined plane, and β is the angle which the line of traction makes with the normal. It is plain, that when β varies, other things being the same, the value of P will vary; there will evidently be some value of β, which will render P the least possible; the direction of P in this case, is called the line of least traction; and it is along this line that a force can be applied with greatest advantage, to draw a body up an inclined plane. If we examine the expression for P, in Equation (51), we see that the numerator remains constant; therefore, the expression for P will be least possible when the denominator is the greatest possible. By a simple process of the Differential Calculus, it may be shown that the denominator will be the greatest possible, or a maximum, when,

$$f = \cot \beta, \text{ or } f = \tan(90° - \beta).$$

That is, the power will be applied most advantageously, when it makes an angle with the inclined plane equal to the angle of friction.

From the second value of P, it may be shown, in like manner, that a force will be most advantageously applied, to prevent a body from sliding down the plane, when its direction makes an angle with the plane equal to the supplement of the angle of friction, the angle being estimated as before from that part of the plane lying above the body.

Friction on an Axle.

110. Let it be required to determine the position of equilibrium of a horizontal axle, resting in a cylindrical box, when the power is just on the point of overcoming the friction between the axle and box.

Let O' be the centre of a cross section of the axle, O the centre of the cross section of the box, and N their point of contact, when the power is on the point of overcoming the friction between the axle and box. The element through N will be the line of contact of the axle and box.

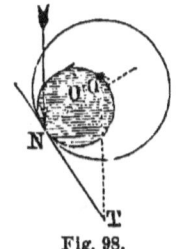

Fig. 98.

When the axle is only acted upon by its own weight, the element of contact will be the lowest element of the box. If, now, a power be applied to turn the axle in the direction indicated by the arrow-head, the axle will roll up the inside of the box until the resultant of all the forces acting upon it becomes normal to the surface of the axle at some point of the element through N. This normal force pressing the axle against the box, will give rise to a force of friction acting tangentially upon the axle, which will be exactly equal to the tangential force applied at the circumference of the axle to produce rotation. If the axle be rolled further up the side of the box, it will slide back to N; if it be moved down the box, it will roll back to N, under the action of the force. In this position of the axle, it is in the condition of a body resting upon an inclined plane, just on the point of sliding down the plane, but restrained by the force of friction. Hence, if a plane be passed tangent to the surface of the box, along the element N, it will make with the horizon an angle equal to the angle of friction. The relation between the power and resistance may then be found, as in Art. 108.

CHAPTER V.

RECTILINEAR AND PERIODIC MOTION.

Motion.

111. A material point is in motion when it continually changes its position in space. When the path of the moving point is a straight line, the motion is *rectilinear;* when it is a curved line, the motion is *curvilinear*. When the motion is curvilinear, we may regard the path as made up of infinitely short straight lines; that is, we may consider it as a polygon, whose sides are infinitely small. If any side of this polygon be prolonged in the direction of the motion, it will be a tangent to the curve. Hence, we say, that *a point always moves in the direction of a tangent to its path*.

Uniform Motion.

112. UNIFORM MOTION is that in which the moving point describes equal spaces in any arbitrary equal portions of time. If we denote the space described in one second by v, and the space described in t seconds by s, we shall have, from the definition,

$$s = vt; \quad \therefore \; v = \frac{s}{t} \quad \dots \quad (55.)$$

From the first of these equations, we see that *the space described in any time is equal to the product of velocity and the time;* and, from the second, we see that *the velocity is equal to the space described in any time, divided by that time.*

These laws hold true for all cases of uniform motion. If we denote by ds the space described in the infinitely short time dt, we shall have, from the last principle,

$$v = \frac{ds}{dt} \quad \dots \dots \dots \quad (56.)$$

which is the differential equation of uniform motion, v being constant. Clearing this equation of fractions, and integrating, we have,
$$s = vt + C \quad \ldots \quad (57.)$$
which is the most general equation of uniform motion. If, in (57), we make $t = 0$, we shall have,
$$s = C.$$

Hence, we see that the constant of integration represents the space passed over by the point, from the origin of spaces up to the beginning of the time t. This space is called the *initial space*. Denoting it by s', we have,
$$s = vt + s' \quad \ldots \quad (58.)$$

If $s' = 0$, the origin of spaces corresponds to the origin of times, and we have,
$$s = vt,$$
the same as the first of Equations (55.)

Varied Motion.

113. VARIED MOTION is that in which the velocity is continually changing. It can only result from the action of an incessant force.

To find the differential equations of varied motion, let us denote the velocity at the time t, by v, and the space passed over up to that time, by s. In the succeeding instant dt, the space described will be ds, and the velocity generated will be dv. Now, the space ds, which is described in the infinitely small time dt, may be regarded as having been described with the uniform velocity v. Hence, from Equation (55), we have,
$$v = \frac{ds}{dt} \quad \ldots \quad (59.)$$

Let us denote the acceleration due to the incessant force at the time t, by φ. We have seen (Art. 24), that the measure of the acceleration due to a force, is the velocity that it can impart in a unit of time, on the hypothesis that it acts uniformly during that time. Now, it is plain that, so long as the force acts uniformly, the velocity generated will be proportional to the time, and, consequently, the measure of the acceleration will be, *the quotient obtained by dividing the velocity generated in any time, by that time.* The quantity φ is, in general, variable; but it may be regarded as constant during the instant dt; and from what has just been said, we shall have,

$$\varphi = \frac{dv}{dt} \quad \ldots \quad \ldots \quad (60.)$$

Differentiating Equation (59), we have,

$$dv = \frac{d^2s}{dt};$$

which, being substituted in Equation (60) gives,

$$\varphi = \frac{d^2s}{dt^2} \quad \ldots \quad \ldots \quad (61.)$$

Equations (59), (60), and (61) are the differential equations required. The acceleration φ, is the measure of the force exerted when the mass moved is the unit of mass (Art. 24); in any other case, it must be multiplied by the mass. Denoting the entire moving force applied to the mass m by F, we shall have,

$$F = m\varphi = m\frac{d^2s}{dt^2} \quad \ldots \quad (62.)$$

This value of F is the measure of the effective moving force in the direction of the body's motion. When a body moves upon any curve in space, the motion may be regarded as taking place in the direction of three rectangular axes. If we denote the effective components of the moving force in the direction of these axes, by X, Y, and Z, the spaces described being denoted by x, y, and z, we shall have, from (62),

$$X = m\frac{d^2x}{dt^2}, \quad Y = m\frac{d^2y}{dt^2}, \quad Z = m\frac{d^2z}{dt^2}.$$

Uniformly Varied Motion.

114. UNIFORMLY VARIED MOTION is that in which the velocity increases or diminishes uniformly. In the former case, the motion is *accelerated ;* in the latter case, it is *retarded.* In both cases, *the moving force is constant.* Denoting the acceleration due to this constant force, by f, we shall have, from Equation (61),

$$\frac{d^2s}{dt^2} = f \quad \ldots \quad (63.)$$

Multiplying by dt, and integrating, we have,

$$\frac{ds}{dt} = ft + C \quad \ldots \quad (64.)$$

or, since $\frac{ds}{dt}$ is equal to v, Equation (59),

$$v = ft + C \quad \ldots \quad (65.)$$

Multiplying both members of (64) by dt, and integrating, we have,

$$s = \tfrac{1}{2}ft^2 + Ct + C' \quad \ldots \quad (66.)$$

Equations (65) and (66) express the relations between the velocity, space, and time, in the most general case of uniformly varied motion. These equations involve the two constants of integration C and C', which serve to make them conform to the different cases that may arise. To determine the value of these constants, make $t = 0$ in the two equations, and denote the corresponding values of v and s, by v' and s'. We shall have,

$$C = v'.$$

$$C' = s'.$$

That is, C is equal to the velocity at the beginning of the time t, and C' is equal to space passed over up to the same time. These values of the velocity and space are called, respectively, *the initial velocity*, and *the initial space*. Substituting for C and C' these values in (65) and (66), they become,

$$v = v' + ft \quad \ldots \quad (67.)$$

$$s = s' + v't + \tfrac{1}{2}ft^2 \quad \ldots \quad (68.)$$

From these equations, we see that the velocity at any time t, is made up of two parts, *the initial velocity*, and *the velocity generated during the time t;* we also see, that the space is made up of three parts, *the initial space, the space due to the initial velocity for the time t,* and *the space due to the action of the incessant force during the same time.*

By giving suitable values to v' and s', Equations (67) and (68) may be made to express every phenomenon of varied motion. If we suppose both v' and s' equal to 0, the body will move from a state of rest at the origin of times, and Equations (67) and (68) will become,

$$v = ft \quad \ldots \quad (69.)$$

$$s = \tfrac{1}{2}ft^2 \quad \ldots \quad (70.)$$

From the first of these equations, we see that, in uniformly varied motion, *the velocity varies as the time;* and, from the second one, we see that *the space described varies as the square of the time.*

If, in Equation (70), we make $t = 1$, we have,

$$s = \tfrac{1}{2}f; \quad \text{or,} \quad f = 2s.$$

That is, when a body moves from a state of rest, under the action of a constant force, the acceleration is equal to twice the space passed over in the first second of time.

If, in the preceding equations, we suppose f to be essentially positive, the motion will be *uniformly accelerated;* if we suppose it to be negative, the motion will be *uniformly retarded.* In the latter case, Equations (67) and (68) become,

$$v = v' - ft \quad \ldots \ldots \quad (71.)$$

$$s = s' + v't - \tfrac{1}{2}ft^2 \quad \ldots \quad (72.)$$

Application to Falling Bodies.

115. THE FORCE OF GRAVITY is the force exerted by the earth upon all bodies exterior to it, tending to draw them towards it. It is found by observation, that *this force is directed towards the centre of the earth*, and that *its intensity varies inversely, as the square of the distance from the centre.*

Since the centre of the earth is so far distant from the surface, the variation in intensity for small elevations above the surface will be inappreciable. Hence, we may regard the force of gravity at any place on the earth's surface, and for small elevations at that place, as constant, in which case, the equations of the preceding article become immediately applicable. The force of gravity acts equally upon all the particles of a body, and were there no resistance offered, it would impart the same velocity, in the same time, to any two bodies whatever. The atmosphere is a cause of resistance, tending to retard the motion of all bodies falling through it; and of two bodies of equal mass, it retards that one the most, which offers the greatest surface to the direction of the motion. In discussing the laws of falling bodies, it will, therefore, be found convenient, in the first place, to regard them as being situated in vacuum, after which, a method will be pointed out, by means of which the velocities can be so diminished, that atmospheric resistance may be neglected.

Let us denote the acceleration due to gravity, at any point on the earth's surface, by g, and the space fallen through in the time t, by h. Then, if the body moves from a state of rest at the origin of times, Equations (69) and (70) will give,

$$v = gt \quad \ldots \ldots \quad (73.)$$

$$h = \tfrac{1}{2}gt^2 \quad \ldots \ldots \quad (74.)$$

RECTILINEAR MOTION. 149

From these equations, we see that *the velocities at two different times are proportional to the times, and the spaces to the squares of the times.*

It has been found by experiment that the velocity imparted to a body in one second of time by the action of the force of gravity in the latitude of New York, is about $32\frac{1}{6}$ feet. Making $g = 32\frac{1}{6}$ ft., and giving to t the successive values 1^s, 2^s, 3^s, &c., in Equations (73) and (74), we shall have the results indicated in the following

TABLE.

TIME ELAPSED.	VELOCITIES ACQUIRED.	SPACES DESCRIBED.
SECONDS.	FEET.	FEET.
1	$32\frac{1}{6}$	$16\frac{1}{12}$
2	$64\frac{1}{3}$	$64\frac{1}{3}$
3	$96\frac{1}{2}$	$144\frac{3}{4}$
4	$128\frac{2}{3}$	$257\frac{1}{3}$
5	$150\frac{5}{6}$	$402\frac{1}{12}$
&c.	&c.	&c.

Solving Equation (74) with respect to t, we have,

$$t = \sqrt{\frac{2h}{g}} \quad \ldots \ldots \quad (75.)$$

That is, *the time required for a body to fall through any height is equal to the square root of the quotient obtained by dividing twice the height in feet by* $32\frac{1}{6}$.

Substituting this value of t in Equation (73), we have.

$$v = g\sqrt{\frac{2h}{g}}, \quad \text{or} \quad v^2 = 2gh;$$

whence, by solving with reference to v and h respectively,

$$v = \sqrt{2gh}, \quad \text{and} \quad h = \frac{v^2}{2g} \quad . \quad . \quad (76.)$$

These equations are of frequent use in dynamical investigations. In them the quantity v is called *the velocity due to the height h*, and the quantity h, *the height due to the velocity v*.

If we suppose the body to be projected downwards with a velocity v', the circumstances of motion will be made known by the Equations,

$$v = v' + gt,$$

$$h = v't + \tfrac{1}{2}gt^2.$$

In these equations we have supposed the origin of spaces to be at the point at which the body is projected downwards.

Motion of Bodies projected vertically upwards.

116. Suppose a body to be projected vertically upwards from the origin of spaces with a velocity v', and afterwards to be acted upon by the force of gravity. In this case, the force of gravity acts to retard the motion. Making in (71) and (72), $s' = o$, $f = g$, and $s = h$, they become,

$$v = v' - gt \quad . \quad . \quad . \quad . \quad . \quad (77.)$$

$$h = v't - \tfrac{1}{2}gt^2 \quad . \quad . \quad . \quad . \quad (78.)$$

In these equations, h is positive when estimated upwards from the origin of spaces, and consequently negative, when estimated downwards from the same point.

From Equation (77), we see that the velocity diminishes as the time increases. The velocity will be 0, when,

$$v' - gt = 0, \quad \text{or when} \quad t = \frac{v'}{g}.$$

If t continues to increase beyond the value $\dfrac{v'}{g}$, v will become negative, and the body will retrace its path. Hence, *the time required for the body to reach its highest elevation, is equal to the initial velocity divided by the force of gravity.*

Eliminating t from Equations (77) and (78), we have,

$$h = \frac{v'^2 - v^2}{2g} \quad \ldots \quad (79.)$$

Making $v = 0$, in the last equation, we have,

$$h = \frac{v'^2}{2g} \quad \ldots \quad (80.)$$

Hence, *the greatest height to which the body will ascend, is equal to the square of the initial velocity, divided by twice the force of gravity.*

This height is that due to the initial velocity (Art. 115).

If, in Equation (77), we make $t = \dfrac{v'}{g} - t'$, we find,

$$v = gt' \quad \ldots \quad (81.)$$

If, in the same equation, we make $t = \dfrac{v'}{g} + t'$, we find,

$$v = -gt' \quad \ldots \quad (82.)$$

Hence, *the velocities at equal times before and after reaching the highest points, are equal.*

The difference of signs shows that the body is moving in opposite directions at the times considered.

If we substitute these values of v successively, in Equation (79), we shall, in both cases, find

$$h = \frac{v'^2 - g^2 t'^2}{2g};$$

which shows that the points at which the velocities are equal, both in ascending and descending, are equally distant from the highest point; that is, they are coincident. Hence, *if a body be projected vertically upwards, it will ascend to a certain point, and then return upon its path, in such a manner, that the velocities in ascending and descending will be equal at the same points.*

EXAMPLES.

1. Through what distance will a body fall from a state of rest in vacuum, in 10 seconds, and through what space will it fall during the last second ? *Ans.* $1608\frac{1}{3}$ ft., and $305\frac{1}{2}$ ft.

2. In what time will a body fall from a state of rest through a distance of 1200 feet ? *Ans.* 8.63 sec.

3. A body was observed to fall through a height of 100 feet in the last second. How long was the body falling, and through what distance did it descend ?

SOLUTION.

If we denote the distance by h, and the time by t, we shall have,

$$h = \tfrac{1}{2}gt^2, \text{ and } h - 100 = \tfrac{1}{2}g(t-1)^2 ;$$

$$\therefore\ t = 3.6 \text{ sec.,} \quad \text{and} \quad h = 208.44 \text{ ft. } Ans.$$

4. A body falls through a height of 300 feet. Through what distance does it fall in the last two seconds?

The entire time occupied, is 4.32 sec. The distance fallen through in 2.32 sec., is 86.57 ft. Hence, the distance required is 300 ft. − 86.57 ft. = 213.43 ft. *Ans.*

5. A body is projected vertically upwards, with a velocity of 60 feet. To what height will it rise ? *Ans.* 55.9 ft.

6. A body is projected vertically upwards, with a velocity of 483 ft. In what time will it rise to a height of 1610 feet ?

We have, from Equation (78),

$$1610 = 483t - 16\tfrac{1}{12}t^2 ; \quad \therefore\ t = \tfrac{2898}{193} \pm \tfrac{2161}{193} ;$$

$$\text{or, } t = 26.2 \text{ sec., and } t = 3.82 \text{ sec.}$$

The smaller value of t gives the time required ; the larger value of t gives the time occupied in rising to its greatest height, and returning to the point which is 1610 feet from the starting point.

7. A body is projected vertically upwards, with a velocity of 161 feet, from a point $214\frac{2}{3}$ feet above the earth. In what time will it reach the surface of the earth, and with what velocity will it strike?

SOLUTION.

The body will rise from the starting point 402.9 ft. The time of rising will be 5 sec.; the time of falling from the highest point to the earth will be 6.2 sec. Hence, the required time is 11.2 sec. The required velocity is 199 ft.

8. Suppose a body to have fallen through 50 feet, when a second begins to fall just 100 feet below it. How far will the latter body fall before it is overtaken by the former?

Ans. 50 feet

Restrained Vertical Motion.

117. We have seen that the entire force exerted in moving a body is equal to the *acceleration*, multiplied by the *mass* (Art. 24). Hence, the acceleration is equal to the *moving force*, divided by the mass. In the case of a falling body, the moving force varies directly as the mass moved; and, consequently, the acceleration is independent of the mass. If, by any combination, the moving force can be diminished whilst the mass remains unchanged, there will be a corresponding diminution in the acceleration. This object may be obtained by the combination represented in the figure. A represents a fixed pulley, mounted on a horizontal axis, in such a manner that the friction shall be as small as possible; W and W' are unequal weights, attached to a flexible cord passing over the pulley. If we suppose the weight W greater than W', the former will descend and draw the latter up. If the difference is very small, the motion will be very slow, and if the instrument is nicely constructed,

Fig. 99.

we may neglect all hurtful resistances as inappreciable. Denote the masses of the weights W and W', by m and m', and the force of gravity, by g. The weight W is urged downwards by the moving force mg, and this motion is resisted by the moving force $m'g$. Hence, the entire moving force is equal to $mg - m'g$, or, $(m - m')g$, and the entire mass moved, is $m + m'$, since the cord joining the weights is supposed inextensible. If we denote the *acceleration* by g', we shall have, from what was said at the beginning of this article,

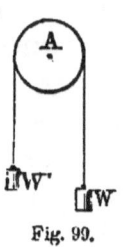

Fig. 90.

$$g' = \frac{m - m'}{m + m'} g \quad . \quad . \quad . \quad . \quad (83.)$$

By diminishing the difference between m and m', we may make the acceleration as small as we please. It is plain that g' is constant; hence, the motion of W is *uniformly varied*. If we replace g by $\dfrac{m - m'}{m + m'} g$, in Equations (73) and (74), they will make known the circumstances of motion of the body W. This principle is employed to illustrate the laws of falling bodies by means of ATWOOD's machine.

Had the two weights under consideration been attached to the extremities of cords passing around a wheel and its axle, and in different directions, it might have been shown that the motion would be uniformly varied, when the moment of either weight exceeded that of the other. The same principle holds good in the more complex combinations of pulleys, wheels and axles, &c. In practice, however, the hurtful resistances increase so rapidly, that even when the moving force remains constant, the velocity soon attains a maximum limit, after which the motion will be sensibly uniform.

EXAMPLES.

1. Two weights of 5 lbs. and 4 lbs., respectively, are suspended from the extremities of a cord passing over a

fixed pulley. What distance will each weight describe in the first second of time, what velocity will be generated in one second, and what will be the tension of the connecting cord?

SOLUTION.

Since the masses are proportional to the weights, we shall have,

$$g' = \frac{5-4}{5+4}g = \frac{1}{9} \times 32\frac{1}{6} \text{ ft.} = 3.574 \text{ ft.}$$

Hence, the velocity generated is 3.574 ft., and the space passed over is 1.787 ft. To find the tension of the string, denote it by x. The moving force acting upon the heavier body, is $(5-x)g$, and the acceleration due to this force, $\left(\frac{5-x}{5}\right)g$; the moving force acting upon the lighter body, is $(x-4)g$, and the corresponding acceleration, $\left(\frac{x-4}{4}\right)g$. But since the two bodies move together, these accelerations must be equal. Hence,

$$\left(\frac{5-x}{5}\right)g = \left(\frac{x-4}{4}\right)g;$$

$\therefore x = 4\frac{4}{9}$ lbs., the required tension.

2. A weight of 1 lb., hanging on a pulley, descends and drags a second weight of 5 lbs. along a horizontal plane. Neglecting hurtful resistances, to what will the accelerating force be equal, and through what space will the descending body move in the first second?

SOLUTION.

The moving force is equal to $1 \times g$, and the mass moved is equal to 6. Hence, the acceleration is equal to $\frac{g}{6} = 5.3622$ ft., and the space described will be equal to 2.6811 ft.

3. Two bodies, each weighing 5 lbs., are attached to a string passing over a fixed pulley. What distance will each

body move in 10 seconds, when a pound weight is added to one of them, and what velocity will have been generated at the end of that time?

SOLUTION.

The acceleration will be equal to $\frac{1}{11}g = 2.924$ ft. $= g'$. But, $s = \tfrac{1}{2}g't^2$, $v = g't$. Hence, the space described in 10 seconds is 146.2 ft., and the velocity generated is 29.24 ft.

4. Two weights, of 16 oz. each, are attached to the ends of a string passing over a fixed pulley. What weight must be added to one of them, that it may descend through a foot in two seconds?

SOLUTION.

Denote the required weight by x; the acceleration will be equal to $\dfrac{x}{32 + x} g = g'$. But $s = \tfrac{1}{2}g't^2$: making $s = 1$ and $t = 2$, we have,

$$1 = \frac{2x}{32 + x} \times 32\tfrac{1}{6}; \quad \therefore \ x = 0.505 \text{ oz. } Ans.$$

Atwood's Machine.

118. Atwood's machine is a contrivance to illustrate the laws of falling bodies. It consists of a vertical post AB, about 12 feet in height, supporting, at its upper extremity, a fixed pulley A. To obviate, as much as possible, the resistance of friction, the axle is made to turn upon friction rollers. A fine silk string passes over the pulley, and at its two extremities are fastened two equal weights C and D. In order to impart motion to the weights, a small weight G, in the form of a bar, is laid upon the weight C, and by diminishing its mass, the acceleration may be rendered as small as desirable. The vertical rod AB, graduated to feet and decimals, is provided with two sliding stages E and F; the upper one is in the form of a ring, which will permit the

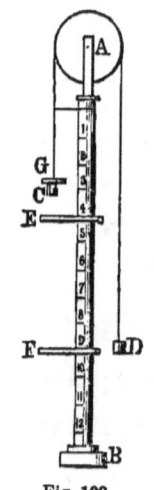

Fig. 100.

weight C, to pass, but not the bar G; the lower one is in the form of a plate, which is intended to intercept the weight C. There is also connected with the instrument a seconds pendulum for measuring time.

Let us suppose that the weights of C and D, are each equal to 181 grains, and that the weight of the bar G, is 24 grains. Then will the acceleration be

$$g' = \frac{24}{362 + 24} g = 2 \text{ ft.};$$

and since $h = \tfrac{1}{2}g't^2$, and $v = g't$ (Art. 115), we shall have, for the case in question,

$$h = t^2, \text{ and } v = 2t.$$

If, in these equations, we make $t = 1$ sec., we shall have $h = 1$, and $v = 2$. If we make $t = 2$ sec., we shall, in like manner, have $h = 4$, and $v = 4$. If we make $t = 3$ sec., we shall have $h = 9$, and $v = 6$, and so on. To verify these results experimentally, commencing with the first. The weight C is drawn up till it comes opposite the 0 of the graduated scale, and the bar G is placed upon it. The weight thus set is held in its place by a spring. The ring E is set at 1 foot from the 0, and the stage F, is set at 3 feet from the 0. When the pendulum reaches one of its extreme limits, the spring is pressed back, the weight C, G descends, and as the pendulum completes its vibration, the bar G strikes the ring, and is retained. The acceleration then becomes 0, and the weight C moves on uniformly, with the velocity that it had acquired, in the first second; and it will be observed that the weight C strikes the second stage just as the pendulum completes its second vibration. Had the stage F been set at 5 feet from the 0, the weight C would have reached it at the end of the third vibration of the pendulum. Had it been 7 feet from the 0, it would have reached it at the end of the fourth vibration, and so on.

To verify the next result, we set the ring E at four feet

from the O, and the stage F at 8 feet from the O, and proceed as before. The ring will intercept the bar at the end of the first vibration, and the weight will strike the stage at the end of the second vibration, and so on.

By making the weight of the bar less than 24 grains, the acceleration is diminished, and, consequently, the spaces and velocities correspondingly diminished. The results may be verified as before.

Motion of Bodies on Inclined Planes.

119. If a body be placed on an inclined plane, and abandoned to the action of its own weight, it will either slide or roll down the plane, provided there be no friction between it and the plane. If the body is spherical, it will roll, and in this case the friction may be disregarded. Let the weight of the body be resolved into two components; one perpendicular to the plane, and the other parallel to it. The plane of these components will be vertical, and it will also be perpendicular to the given plane. The effect of the first component will be counteracted by the resistance of the plane, whilst the second component will act as a constant force, continually urging the body down the plane. The force being constant, the body will have a uniformly varied motion, and Equations (67) and (68) will be immediately applicable. The acceleration will be found by projecting the acceleration due to gravity upon the inclined plane.

Let AB represent a section of the inclined plane made by a vertical plane taken perpendicular to the given plane, and let P be the centre of gravity of a body resting on the given plane. Let PQ represent the acceleration due to gravity, denoted by g, and let PR be the component of g, which is parallel to

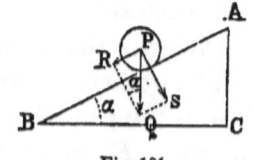

Fig. 101.

AB, denoted by g', PS being the normal component. Denote the angle that AB makes with the horizontal plane by α. Then, since PQ is perpendicular to BC, and QR to

AB, the angle RQP is equal to ABC, or to α. Hence we have, from the right-angled triangle PQR,

$$g' = g\sin\alpha.$$

But the triangle ABC is right-angled, and, if we denote its height AC by h, and its length AB by l, we shall have $\sin\alpha = \dfrac{h}{l}$, which, being substituted above, gives,

$$g' = \frac{gh}{l} \quad \dotfill \quad (84.)$$

This value of g' is the value of the acceleration due to the moving force. Substituting it for f in Equations (67) and (68), we have,

$$v = v' + \frac{gh}{l}t,$$

$$s = s' + v't + \frac{gh}{2l}t^2.$$

If the body starts from rest at A, taken as the origin of spaces, then will $v' = 0$ and $s' = 0$, giving,

$$v = \frac{gh}{l}t. \quad \dotfill \quad (85.)$$

$$s = \frac{gh}{2l}t^2. \quad \dotfill \quad (86.)$$

To find the time required for a body to move from the top to the bottom of the plane, make $s = l$, in (86); there will result,

$$l = \frac{gh}{2l}t^2; \quad \therefore \quad t = l\sqrt{\frac{2}{gh}}. \quad (87.)$$

Hence *the time varies directly as the length, and inversely as the square root of the height.*

For two planes having the same height, but different lengths, the radical factor of the value of t will remain constant. Hence, *the times required for a body to move down any two planes having the same height, are to each other as their lengths.*

To determine the velocity with which a body reaches the bottom of the plane, substitute for t, in Equation (85) *its value taken from Equation* (86). We shall have, after reduction,

$$v = \sqrt{2gh}.$$

But this is the velocity due to the height h (Art. 115). Hence, *the velocity generated in a body whilst moving down any inclined plane, is equal to that generated in falling freely through the height of the plane.*

EXAMPLES.

1. An inclined plane is 10 feet long and 1 foot high. How long will it take for a body to move from the top to the bottom, and what velocity will it acquire in the descent?

SOLUTION.

We have, from Equation (87),

$$t = l\sqrt{\frac{2}{gh}};$$

substituting for l its value 10, and for h its value 1, we have,

$$t = 2\tfrac{1}{2} \text{ seconds nearly.}$$

From the formula $v = \sqrt{2gh}$, we have, by making $h = 1$,

$$v = \sqrt{64.33} = 8.02 \text{ ft.}$$

2. How far will a body descend from rest in 4 seconds, on an inclined plane whose length is 400 feet, and whose height is 300 feet? *Ans.* 193 ft.

3. How long will it take for a body to descend 100 feet on a plane whose length is 150 feet, and whose height is 60 feet? *Ans.* 3.9 sec

4. There is an inclined railroad track, 2½ miles in length, whose inclination is 1 in 35. What velocity will a car attain, in running the whole length of the road, by its own weight, hurtful resistances being neglected?

Ans. 155.75 ft., or, 106.2 m. per hour.

5. A railway train, having a velocity of 45 miles per hour, is detached from the locomotive on an ascending grade of 1 in 200. How far, and for what time, will the train continue to ascend the inclined plane?

SOLUTION.

We find the velocity to be 66 ft. per second. Hence, $66 = \sqrt{2gh}$; or, $h = 67.7$ ft. for the vertical height. Hence, $67.7 \times 200 = 13{,}540$ ft., or, 2.5644 m., the distance which the train will proceed. We have,

$$t = l\sqrt{\frac{2}{gh}} = 410.3 \text{ sec., or, 6 min. 50.3 sec.,}$$

for the time required to come to rest.

6. A body weighing 5 lbs. descends vertically, and draws a weight of 6 lbs. up an inclined plane of 45°. How far will the first body descend in 10 seconds?

SOLUTION.

The moving force is equal to $5 - 6 \sin 45°$; and, consequently, the acceleration,

$$g' = \frac{5 - 6 \sin 45°}{6 + 5} = \frac{.757}{11} = .068818;$$

$$\therefore \quad s = \tfrac{1}{2} g' t^2 = 3.4409 \text{ ft.} \quad Ans.$$

Motion of a Body down a succession of Inclined Planes.

120. If a body start from the top of an inclined plane, with an initial velocity v', it will reach the bottom with a velocity equal to the initial velocity, increased by that due to the height of the plane. This velocity, called the *terminal velocity*, will, therefore, be equal to that which the body would have acquired by falling freely through a height equal to that due to the initial velocity, increased by that of the plane. Hence, if a body start from a state of rest at A, and, after having passed over one inclined plane AB, enters upon a second plane BC, without loss of velocity, it will reach the bottom of the second plane with the same velocity that it would have acquired by falling freely through DC, the sum of the heights of the two planes. Were there a succession of inclined planes, so arranged that there would be no loss of velocity in passing from one to another, it might be shown, by a similar course of reasoning, that the terminal velocity would be equal to that due to the vertical distance of the terminal point below the point of starting.

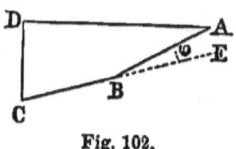

Fig. 102.

By a course of reasoning entirely analagous to that employed in discussing the laws of motion of bodies projected vertically upwards, it might be shown that, if a body were projected upwards, in the direction of the lower plane, with the terminal velocity, it would ascend along the several planes to the top of the highest one, where the velocity would be reduced to 0. The body would then, under the action of its own weight, retrace its path in such a manner that the velocity at every point in descending would be the same as in ascending, but in a contrary direction. The time occupied by the body in passing over any part of its path in descending, would be exactly equal to that occupied in passing over the same portion in ascending.

In the preceding discussion, we have supposed that there is no loss of velocity in passing from one plane to another. To ascertain under what circumstances this condition will be fulfilled, let us take the two planes AB and BC. Prolong BC upwards, and denote the angle ABE, by φ. Denote the velocity of the body on reaching B, by v'. Let v' be resolved into two components, one in the direction of BC, and the other at right angles to it. The effect of the latter will be destroyed by the resistance of the plane, and the former will be the effective velocity in the direction of the plane BC. From the rule for decomposition of velocities, we have, for the effective component of v', the value $v' \cos\varphi$. Hence, the loss of velocity due to change of direction, is $v' - v' \cos\varphi$; or, $v'(1 - \cos\varphi)$, which is equal to v' ver-sinφ. But when φ is infinitely small, its versed-sine is 0, and there will be no loss of velocity. Hence, the loss of velocity due to change of direction will always be 0, when the path of the body is a curved line. This principle is general, and may be enunciated as follows: *When a body is constrained to describe a curvilinear path, there will be no loss of velocity in consequence of the change in direction of the body's motion.*

Periodic Motion.

121. *Periodic motion* is a kind of variable motion, in which the spaces described in certain equal periods of time are equal. This kind of motion is exemplified in the phenomena of *vibration*, of which there are two cases.

1st. *Rectilinear vibration.* Theory indicates, and experiment confirms the fact, that if a particle of an elastic fluid be slightly disturbed from its place of rest, and then abandoned, it will be urged back by a force, varying *directly* as its distance from the position of equilibrium; on reaching this position, the particle will, by virtue of its inertia, pass to the other side, again to be urged back, and so on. To determine the time required for the particle to pass from one extreme position to the opposite one and back, let us denote the displacement at any time t by s, and the acceleration due to the *restoring force* by φ; then, from the law of the force, we shall have $\varphi = n^2 s$, in which n is constant for the same fluid at the same temperature. Substituting for φ its value, Equation (61), and recollecting that φ acts in a direction contrary to that in which s is estimated, we have,

$$-\frac{d^2 s}{dt^2} = n^2 s$$

Multiplying both members by $2ds$, we have,

$$-\frac{2ds\,d^2s}{dt^2} = 2n^2 s\,ds;$$

whence, by integration,

$$-\frac{ds^2}{dt^2} = n^2 s^2 + C = -v^2.$$

The velocity v will be 0 when s is greatest possible; denoting this value of s by a, we shall have,

$$n^2 a^2 + C = 0; \quad \text{whence,} \quad C = -n^2 a^2.$$

Substituting this value of C in the preceding equation, it becomes,

$$v^2 = \frac{ds^2}{dt^2} = n^2(a^2 - s^2); \quad \text{whence,} \quad n\,dt = \frac{ds}{\sqrt{a^2 - s^2}}. \quad (88.)$$

Integrating the last equation, we have,

$$nt + C = \sin^{-1}\frac{s}{a} \quad \ldots \quad \ldots \quad (89.)$$

Taking the integral between the limits $s = +a$ and $s = -a$, and denoting the corresponding time by $\tfrac{1}{2}\tau$, τ being the time of a double vibration, we have,

$$\tfrac{1}{2}n\tau = \pi; \quad \text{whence,} \quad \tau = \frac{2\pi}{n}.$$

The value of τ is independent of the extent of the excursion, and dependent only upon n. Hence, in the same medium, and at the same temperature, the time of vibration is *constant*.

These principles are of utility in discussing the subjects of sound, light, &c.

2ndly. *Curvilinear vibration.* Let ABC be a vertical plane curve, symmetrical with respect to DB. Let AC be a horizontal line, and denote the distance EB by h. If a body were placed at A and abandoned to the action of its own weight, being constrained to remain on the curve, it would, in accordance with the principles of the last article, move towards B with an accelerated motion, and, on arriving at B, would possess a velocity due to the height h. By virtue of its inertia, it would ascend the branch BC with a retarded motion, and would finally reach C, where its velocity would be 0. The body would then be in the same condition that it was at A, and would, consequently, descend to B and again ascend to A, whence it would again descend, and so on. Were there no retarding causes, the motion would continue for ever. From what has preceded, it follows that the time occupied by the body in passing from A to B is equal to that in passing from B to C, and also the time in passing from C to B is equal to that in passing from B to A. Further, the velocities of the body when at G and H, any two points lying on the same horizontal, are equal, either being that due to the height EK. These principles are of utility in discussing the pendulum.

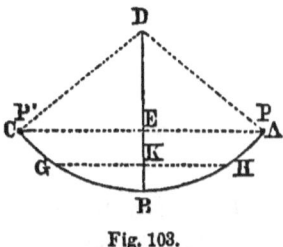

Fig. 103.

Angular Velocity.

122. When a body revolves about an axis, its points being at different distances from the axis, will have different velocities. The *angular velocity* is the velocity of a point whose distance from the axis is equal to 1. To obtain the velocity of any other point, we multiply its distance from the axis by the angular velocity. To find a general expression for the velocity of any point of a revolving body, let us denote the angular velocity by ω, the space passed over by a point at the unit's distance from the axis in the time dt, by $d\theta$. The quantity $d\theta$ is an infinitely small arc, having a radius equal to 1; and, as in Art. 113, it is plain that we may regard the angular motion as uniform, during the infinitely small time dt. Hence, as in Article 113, we have,

$$\omega = \frac{d\theta}{dt} \quad \ldots \ldots \quad (90.)$$

If we denote the distance of any point from the axis by l, and its velocity by v, we shall have,

$$v = l\omega; \quad \text{or,} \quad v = l\frac{d\theta}{dt} \quad \ldots \quad (91.)$$

The Simple Pendulum.

123. A PENDULUM is a heavy body suspended from a horizontal axis, about which it is free to vibrate. In order to investigate the circumstances of vibration, let us first consider the hypothetical case of a single material point vibrating about an axis, to which it is attached by a rod destitute of weight. Such a pendulum is called a SIMPLE PENDULUM. The laws of vibration, in this case, will be identical with those explained in Art. 121, the arc ABC being the arc of a circle. The motion is, therefore, *periodic*.

Let ABC be the arc through which the vibration takes place, and denote its radius by l. The angle CDA is called the *amplitude* of vibration; half of this angle ADB, denoted by α, is called the *angle of deviation;* and l is called *the length of the pendulum*. If the point starts from rest, at A, it will, on reaching any point H, of its path, have a velocity v, due to the height EK, denoted by h. Hence,

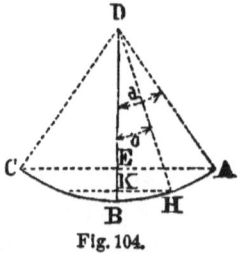

Fig. 104.

$$v = \sqrt{2gh} \quad \ldots \ldots \quad (92.$$

If we denote the variable angle HDB by θ, we shall

have $DK = l\cos\theta$; we shall also have $DE = l\cos\alpha$; and since h is equal to $DK - DE$, we shall have,

$$h = l(\cos\theta - \cos\alpha).$$

Which, being substituted in the preceding formula, gives,

$$v = \sqrt{2gl(\cos\theta - \cos\alpha)}.$$

From the preceding article, we have,

$$v = l\frac{d\theta}{dt}.$$

Equating these two values of v, we have,

$$l\frac{d\theta}{dt} = \sqrt{2gl(\cos\theta - \cos\alpha)}.$$

Whence, by solving with respect to dt,

$$dt = \sqrt{\frac{l}{2g}} \cdot \frac{d\theta}{\sqrt{\cos\theta - \cos\alpha}} \quad \cdots \quad (93.)$$

If we develop $\cos\theta$ and $\cos\alpha$ into series, by McLaurin's theorem, we shall have,

$$\cos\theta = 1 - \frac{\theta^2}{2} + \frac{\theta^4}{1.2.3.4} - \&c.;$$

$$\cos\alpha = 1 - \frac{\alpha^2}{2} + \frac{\alpha^4}{2.3.4} - \&c.$$

When α is very small, say one or two degrees, θ being still smaller, we may neglect all the terms after the second as inappreciable, giving

$$\cos\theta = 1 - \frac{\theta^2}{2};$$

$$\cos\alpha = 1 - \frac{\alpha^2}{2};$$

or, $$\cos\theta - \cos\alpha = \tfrac{1}{2}(\alpha^2 - \theta^2).$$

Substituting in Equation (93), it becomes,

$$dt = \sqrt{\frac{l}{g}} \cdot \frac{d\theta}{\sqrt{\alpha^2 - \theta^2}} \quad \dots \quad (94.)$$

Integrating Equation (94), we have,

$$t = \sqrt{\frac{l}{g}} \sin^{-1} \frac{\theta}{\alpha} + C.$$

Taking the integral between the limits $\theta = -\alpha$, and $\theta = +\alpha$, t will denote the time of one vibration, and we shall have,

$$t = \pi \sqrt{\frac{l}{g}} \quad \dots \quad (95.)$$

Hence, *the time of vibration of a simple pendulum is equal to the number* 3.1416, *multiplied into the square root of the quotient obtained by dividing the length of the pendulum by the force of gravity.*

For a pendulum, whose length is l', we shall have,

$$t' = \pi \sqrt{\frac{l'}{g}} \quad \dots \quad (96.)$$

From Equations (95) and (96), we have, by division,

$$\frac{t}{t'} = \frac{\sqrt{l}}{\sqrt{l'}}; \quad \text{or,} \quad t : t' :: \sqrt{l} : \sqrt{l'} \quad (97.)$$

That is, *the times of vibration of two simple pendulums, are to each other as the square roots of their lengths.*

If we suppose the lengths of two pendulums to be the same, but the force of gravity to vary, as it does slightly in different latitudes, and at different elevations, we shall have,

$$t = \pi \sqrt{\frac{l}{g}}, \quad \text{and} \quad t'' = \pi \sqrt{\frac{l}{g''}}.$$

Whence, by division,

$$\frac{t}{t''} = \sqrt{\frac{g''}{g}}, \quad \text{or,} \quad t : t'' :: \sqrt{g''} : \sqrt{g} \quad . \quad (98.)$$

That is, *the times of vibration of the same simple pendulum, at two different places, are to each other inversely as the square roots of the forces of gravity at the two places.*

If we suppose the times of vibration to be the same, and the force of gravity to vary, the lengths will vary also, and we shall have,

$$t = \pi\sqrt{\frac{l}{g}}, \quad \text{and} \quad t = \pi\sqrt{\frac{l'}{g'}}.$$

Equating these values and squaring, we have,

$$\frac{l}{g} = \frac{l'}{g'}; \quad \text{or,} \quad l : l' :: g : g' \quad . \quad . \quad (99.)$$

That is, *the lengths of simple pendulums which vibrate in equal times at different places, are to each other as the forces of gravity at those places.*

Vibrations of equal duration are called *isochronal*.

The Compound Pendulum.

124. A COMPOUND PENDULUM is a heavy body free to oscillate about a horizontal axis. This axis is called the *axis of suspension*. The straight line drawn from the centre of gravity of the pendulum perpendicular to the axis of suspension is called the *axis of the pendulum*.

In all practical applications, the pendulum is so taken that the plane through the axis of suspension and the centre of gravity divides it symmetrically.

Were the elementary particles of the pendulum entirely disconnected, but constrained to remain at invariable distances from the axis of suspension, we should have a collection of simple pendulums. Those at equal distances from the axis would vibrate in equal times; those unequally distant from it would vibrate in unequal times.

Those particles which are at the same distance from the axis of suspension lie upon the surface of a cylinder, whose axis coincides with the axis of suspension, and we may, without at all affecting the time of vibration, suppose them all to be concentrated at the point in which the cylinder cuts the axis of the pendulum. If we suppose the same to be done for each of the concentric cylinders, we may regard the pendulum as made up of a succession of heavy points, $a, b, \ldots p, k$, lying on the axis, firmly connected with each other and with the point of suspension C. The particles a, b, &c., nearest to C will tend to accelerate the motion of the entire pendulum, whilst those most remote, as p, k, &c., will tend to retard it. There must, therefore, be some intermediate point, as O, which will vibrate precisely as though it were not connected with the system; were the entire mass of the pendulum concentrated at this point it would vibrate in the same time as the given pendulum. This point O is called the centre of oscillation. Hence, the *centre of oscillation of a pendulum* is that point of its axis, at which, if the entire mass of the pendulum were concentrated, its time of vibration would be unchanged. A line drawn through this point, parallel to the axis of suspension, is called *the axis of oscillation*. The distance from the axis of oscillation to the axis of suspension is the length of an *equivalent simple pendulum*, that is, of a simple pendulum, whose time of vibration is the same as that of the compound pendulum.

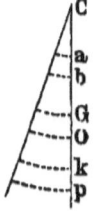

Fig. 105.

To find an expression for CO, C being the axis of suspension, and O the axis of oscillation. Denote CO by l; let G be the centre of gravity, and denote the distance CG by k; denote the masses concentrated at $a, b, \ldots p, k$, by $m, m' \ldots m'', m'''$, and their distances from C by $r, r' \ldots r'', r'''$.

Whatever may be the position of CO, the effective com

ponent of gravity is the same for each particle, and were they free to move, each would have impressed upon it the same velocity that is actually impressed upon O. Denote the angular velocity at any instant, by ω; then will the actual velocity of the mass m, be equal to $r\omega$, and the effective moving force will be equal to $mr\omega$ (Art. 24). Had the mass m been at O, instead of at a, the entire moving force impressed would have been effective, and its measure would have been $ml\omega$. The difference between these forces, or $m(l-r)\omega$, is that portion of the force applied at a which goes to accelerate the motion of the system. The moment of this force with respect to C, is $m(l-r)r\omega$. In like manner, for the force acting at b, which also tends to accelerate the system, we have $m'(l-r')r'\omega$, and so on, for all of the particles between O and C. By a similar course of reasoning, we get, for the moments of the force tending to retard the system, and which are applied at the points p, k, &c., $m''(r''-l)r''\omega$, $m'''(r'''-l)r'''\omega$, &c. But since there is neither acceleration nor retardation, in consequence of the action of these forces, they must be in equilibrium, and, consequently, the sum of the moments of the forces which tend to accelerate the system, must be equal to the sum of the moments, which tend to retard the system. Hence, we have,

$$m(l-r)r\omega + m'(l-r')r'\omega + \&c.$$
$$= m''(r''-l)r''\omega + m'''(r'''-l)r'''\omega + \&c.$$

Striking out the factor ω, and reducing, we have,

$$(mr + m'r' + m''r'' + \&c.)\, l = mr^2 + m'r'^2 + m''r''^2 + \&c.,$$

or, $$\Sigma(mr) \times l = \Sigma(mr^2).$$

Hence,
$$l = \frac{\Sigma(mr^2)}{\Sigma(mr)} \quad \ldots \quad (100.)$$

The expression $\Sigma(mr^2)$, is called the *moment of inertia* of the body with respect to the axis of suspension.

The moment of inertia of a body, with respect to any axis, is the algebraic sum of the products obtained by multiplying the mass of each elementary particle by the square of its distance from the axis.

The expression $\Sigma(mr)$, is called the *moment of the mass*, with respect to the axis of suspension.

The moment of the mass with respect to any axis, is the algebraic sum of the products obtained by multiplying the mass of each elementary particle by its distance from the axis.

From the principle of moments, this is equal to the moment of the entire mass, concentrated at the centre of gravity. Denote the mass, or $\Sigma(m)$, by M, the distance of its centre of gravity from the axis, by k, and we shall have,

$$\Sigma(mr) = Mk \quad \ldots \ldots \quad (101.)$$

Substituting this in Equation (100), we have,

$$l = \frac{\Sigma(mr^2)}{Mk} \quad \ldots \ldots \quad (102.)$$

That is, *the distance from the axis of suspension to the axis of oscillation is equal to the moment of inertia, taken with respect to the axis of suspension, divided by the moment of the mass, taken with respect to the same axis.*

Let the axis of oscillation be taken as an axis of suspension, and denote its distance from the new axis of oscillation by l'. The distances of a, $b \ldots p$, k, from O, will be $l - r$, $l - r'$, &c., and the distance GO will be $l - k$. From the principle just enunciated, we shall have,

$$l' = \frac{\Sigma[m(l-r)^2]}{M(l-k)}.$$

PERIODIC MOTION. 173

Or, performing the operation of squaring and reducing,

$$l' = \frac{\Sigma(ml^2 - 2mrl + mr^2)}{M(l-k)} = \frac{\Sigma(ml^2) - 2\Sigma(mrl) + \Sigma(mr^2)}{M(l-k)}.$$

But l is constant, hence $\Sigma(ml^2) = \Sigma(m) \times l^2 = Ml^2$, also, $2\Sigma(mrl) = 2\Sigma(mr) \times l = 2Mkl$; from Equation (102) we have, $\Sigma(mr^2) = Mkl$. Substituting these values in the preceding equation, we have,

$$l' = \frac{Ml^2 - 2Mkl + Mkl}{M(l-k)} = \frac{M(l-k)l}{M(l-k)}.$$

or,

$$l' = l \quad \ldots \quad (103.)$$

Hence, it follows that *the axes of suspension and oscillation are convertible;* that is, *if either be taken as the axis of suspension, the other will be the axis of oscillation, and the reverse.*

This property of the compound pendulum has been employed to determine experimentally the length of the seconds pendulum, and the value of the force of gravity at different places on the surface of the earth.

A straight bar of iron CD, is provided with two knife-edge axes, A and B, of hardened steel, at right angles to the axis of the bar, and having their edges turned towards each other. These axes are so placed that their plane will pass through the axis of the bar. The pendulum thus constructed is suspended on horixontal plates of polished agate, and allowed to vibrate about each axis in turn till, by filing away one of the ends of the bar, the times of vibration about the two axes are made equal. The distance AB is then equal to the length of the equivalent simple pendulum; that is, of a simple pendulum which will vibrate in the same time as the bar about either axis.

Fig. 106.

To employ the pendulum thus adjusted to find the length of a simple seconds pendulum at any place, the pendulum is carefully suspended, and allowed to vibrate through a very small angle; the number of vibrations is counted, and the time occupied is carefully noted by means of a well-regulated chronometer. The entire time divided by the number of vibrations performed, gives the time of a single vibration. The distance between the axes is carefully measured by an accurate scale of equal parts, which gives the length of the corresponding simple pendulum. To find the length of the simple seconds pendulum, we then make use of Proportion (97), substituting in it for t' and l' the values just found, and for t, 1 second; the only remaining quantity in the proportion is l, which may be found by solving the proportion. This value of l is the required length of the simple seconds pendulum at the place where the observation is made. In making the observations, a variety of precautions must be taken, and several corrections applied, the explanation of which does not fall within the scope of this treatise. It is only intended to point out the general method of proceeding. By a long series of carefully conducted experiments, it has been found that the length of a simple seconds pendulum in the Tower of London is 3.2616 ft., or 39.13921 in. By a similar course of proceeding, the length of the seconds pendulum has been determined for a great number of places on the earth's surface, at different latitudes, and from these results the corresponding values of the force of gravity at those points have been determined according to the following principle:

From Equation (95), which is, $t = \pi\sqrt{\dfrac{l}{g}}$, we find, by solving with respect to g, and making $t = 1$,

$$g = \pi^2 l.$$

From this equation the value of g may be found at different places, by simply substituting for l the length of the seconds pendulum at those places. In this manner, the value of g is found for a great number of places in different latitudes, and from these values the form of the earth's surface may be computed.

It has been ascertained in this manner that if the force of gravity at any point on the earth's surface be denoted by g, the force of gravity at a point whose latitude is 45°, by g', and the latitude of the place where the force of gravity is g', by l, we shall have,

$$g = g'(1 - .002695\cos 2l).$$

PRACTICAL APPLICATIONS OF THE PENDULUM.

125. One of the most important of the applications of the pendulum is to regulating the motion of clocks. A clock consists of a train of wheelwork, the last wheel of the train connecting with the upper extremity of a pendulum-rod by a piece of mechanism called an *escapement*. The wheelwork is maintained in motion by means of a descending weight, or by the elastic force of a coiled spring, and the wheels are so arranged that one tooth of the last wheel in the train escapes from the upper end of the pendulum-rod at each vibration of the pendulum, or at each *beat*. The number of beats is registered and rendered visible on a dial-plate by means of indices, called the *hands of the clock*.

On account of the expansion and contraction of the material of which the pendulum is composed, the length of the pendulum is liable to continual variation, which gives rise to an irregularity in the times of vibration of the pendulum. To obviate this inconvenience, and to render the times of vibration perfectly uniform, several ingenious devices have been resorted to, giving rise to what are called *compensating pendulums*. We shall indicate two of the most important of these combinations, observing that all of the remaining ones are nearly the same in principle, differing only in the modes of application.

Graham's Mercurial Pendulum.

126. GRAHAM's mercurial pendulum consists of a rod of steel about 42 inches long, branched towards its lower end, so as to embrace a cylindrical glass vessel 7 or 8 inches deep, and having 6.8 in. of this depth filled with mercury. The exact quantity of mercury being dependent on the weight and expansibility of the other parts of the pendulum, must be determined by experiment in each individual case When the temperature increases, the steel rod is lengthened, and, at the same time, the mercury expanding, rises in the cylinder. When the temperature decreases, the steel bar is shortened, and the mercury falls in the cylinder. By a proper adjustment of the quantity of mercury, the effect of the lengthening or shortening of the rod is exactly counterbalanced by the rising or falling of the centre of gravity of the mercury, and the axis of oscillation is kept at an invariable distance from the axis of suspension.

Harrison's Gridiron Pendulum.

127. HARRISON's gridiron pendulum consists of five rods of steel and four of brass, placed alternately with each other, the middle rod, or that from which the *bob* is suspended, being of steel. These rods are connected by cross-pieces in such a manner that, whilst the expansion of the steel rods tends to elongate the pendulum, or lower the bob, the expansion of the brass rods tends to shorten the pendulum, or raise the bob. By duly proportioning the sizes and lengths of the bars, the axis of oscillation may be maintained, by the combination, at an invariable distance from the axis of suspension. From what has preceded, it follows that whenever the distance from the axis of oscillation to the axis of suspension remains invariable, the times of vibration must be absolutely equal at the same place. The pendulums just do

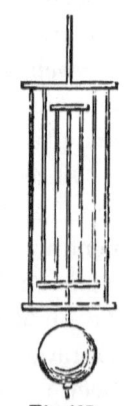

Fig. 107.

scribed are principally used for astronomical clocks, where great accuracy and great uniformity in the measure of time is indispensable.

Basis of a system of Weights and Measures.

128. The pendulum is of further importance, in a practical point of view, in furnishing the standard of comparison which has been made use of as a basis of the English system of weights and measures. The length of the seconds pendulum at any place, can always be found, and it must always be the same at that place. We have seen that this length was determined, with great accuracy, in the Tower of London, to be 3.2616 ft. It has been decreed by the British Government, that the $\frac{1}{3.2616}$th part of the length of the simple seconds pendulum, in the Tower of London, shall be regarded as a *standard foot*. From this, by multiplication and division, every other unit of lineal measure may be derived. By constructing squares and cubes upon the linear units, we at once arrive at the units of area and of volume.

It has further been decreed, that a cubic foot of distilled water, at the temperature of maximum density, shall be regarded as weighing 1000 *standard ounces*. This fixes the ounce; and by multiplication and division, all other units of weight may be derived.

This system enables us to refer to the original standard, when, from any circumstances, doubt may exist as to the accuracy of standard measures. Even should every vestige of a standard be swept from existence, they might be perfectly restored; by the process above indicated.

The American system of weights and measures is adopted from that of Great Britain, and is, in all respects, the same as that above described.

EXAMPLES.

1. The length of a seconds pendulum is 39.13921 in. If it be shortened 0.130464 in., how many vibrations will be gained in a day of 24 hours?

178   MECHANICS.

SOLUTION.

The times of vibration of two pendulums at the same place, are to each other as the square roots of their lengths (Eq. 97). Hence, the number of vibrations made in any given time, are inversely proportional to the square roots of their lengths. If, therefore, we denote the number of vibrations gained in 24 hours, or 86400 seconds, by x, we shall have,

$$86400 : 86400 + x :: \sqrt{39.008747} : \sqrt{39.13921};$$

or, $86400 : 86400 + x :: 6.2457 : 6.2561$.

Whence, $x = 144$, nearly. *Ans.*

2. A seconds pendulum being carried to the top of a mountain, was observed to lose 5 vibrations per day of 86400 seconds. Required the height of the mountain, reckoning the radius of the earth at 4000 miles.

SOLUTION.

The squares of the times of vibration, at any two points, are inversely proportional to the forces of gravity at those points (Eq. 98). But the forces of gravity at the same points are inversely as the squares of their distances from the centre of the earth. Hence, the times of vibration are proportional to the distances of the points from the centre of the earth; and, consequently, the number of vibrations in any given time, as 24 hours, for example, will be inversely as those distances. If, therefore, we denote the height of the mountain in miles by x, we shall have,

$$86400 : 86405 :: 4000 : 4000 + x.$$

Whence, $x = \frac{20000}{86400} = 0.2315$ miles, or, 1222 feet. *Ans.*

3. What is the time of vibration of a pendulum whose length is 60 inches, when the force of gravity is reckoned at $32\frac{1}{6}$ ft? *Ans.* 1.2387 sec.

4. How many vibrations will a pendulum 36 inches in length make in one minute, the force of gravity being the same as before? *Ans.* 62.53.

5. A pendulum is found to make 43170 vibrations in 12 hours. How much must it be shortened that it may beat seconds?

SOLUTION.

We shall have, as in Example 1st,

$$43170 : 43200 :: \sqrt{39.13921} : \sqrt{39.13921 + x}.$$

Whence, $x = 0.0544$ in. *Ans.*

6. In a certain latitude, the length of a pendulum vibrating seconds is 39 inches. What is the length of a pendulum vibrating seconds, in the same latitude, at the height of 21000 feet above the first station, the radius of the earth being 3960 miles? *Ans.* 38.9218 in.

7. If a pendulum make 40000 vibrations in 6 hours, at the level of the sea, how many vibrations will it make in the same time, at an elevation of 10560 feet above the same point, the radius of the earth being 3960 miles?
Ans. 39979.8.

Centre of Percussion.

129. The point O, Fig. 108, is a point at which, if the entire mass were concentrated, and the impressed forces applied to it, the effect produced would be in nowise different from what actually obtains. Were an impulse applied at this point, capable of generating a quantity of motion equal and directly opposed to the resultant of all the quantities of motion of the particles of the body, at any instant, the body would evidently be brought to a state of rest without imparting any shock to the axis of suspension. The direction of the impulse remaining the same as before,

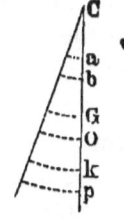

Fig. 108.

no matter what may be its intensity, there will still be no shock on the axis. This point is, therefore, called the *centre of percussion*. We may then define the centre of percussion to be that point of a body restrained by an axis, at which, if the body be struck in a direction perpendicular to a plane passed through this point and the axis of suspension, o shock will be imparted to the axis. It is a matter of common observation that, if a rod held in the hand be struck at a certain point, the hand will not feel the blow, but if it be struck at any other point of its length, there will be a shock felt, the intensity of which will depend upon the intensity of the blow, and upon the distance of its point of application from the first point.

Moment of Inertia.

130. The MOMENT OF INERTIA of a body with respect to an axis, is the *algebraic sum* of the products obtained by multiplying the mass of each elementary particle by the square of its distance from the axis. Denoting the moment of inertia with respect to any axis, by K, the mass of any element of the body, by m, and its distance from the axis, by r, we have, from the definition,

$$K = \Sigma(mr^2) \quad . \quad . \quad . \quad . \quad (104.)$$

The moment of inertia evidently varies, in the same body, according to the position of the axis. To investigate the law of variation, let AB represent any section of the body by a plane perpendicular to the axis; C, the point in which this plane cuts the axis; and G, the point in which it cuts a parallel axis through the centre of gravity. Let P be any element of the body, whose mass is m, and denote PC by r, PG by s, and CG by k.

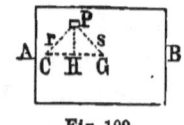

Fig. 109.

From the triangle CPG, according to a principle of Trigonometry, we have,

$$r^2 = s^2 + k^2 - 2sk \cos CGP.$$

Substituting in (104), and separating the terms, we have,

$$K = \Sigma(ms^2) + \Sigma(mk^2) - 2\Sigma(msk\cos CGP).$$

Or, since k is constant, and $\Sigma(m) = M$, the mass of the entire body, we have,

$$K = \Sigma(ms^2) + Mk^2 - 2k\Sigma(ms\cos CGP).$$

But $s\cos CGP = GH$, the lever arm of the mass m, with respect to the axis through the centre of gravity. Hence, $\Sigma(ms \cos CGP)$, is the algebraic sum of the moments of all the particles of the body with respect to the axis through the centre of gravity; but from the principle of moments, this is equal to 0. Hence,

$$K = \Sigma(ms^2) + Mk^2 \quad . \quad . \quad . \quad (105.)$$

The first term of the second member of (105), is the expression for the moment of inertia, with respect to the axis through the centre of gravity.

Hence, *the moment of inertia of a body with respect to any axis, is equal to the moment of inertia with respect to a parallel axis through the centre of gravity,* plus *the mass of the body into the square of the distance between the two axes.*

The moment of inertia is, therefore, the least possible, when the axis passes through the centre of gravity. If any number of parallel axes be taken at equal distances from the centre of gravity, the moment of inertia with respect to each, will be the same.

The moment of inertia of a body with respect to any axis, may be determined experimentally as follows. Make the axis horizontal, and allow the body to vibrate about it, as a compound pendulum. Find the time of a single vibration, and denote it by t. This value of t, in Equation (95), makes known the value of l. Determine the centre of gravity by some one of the methods given, and denote its distance from the axis, by k. Find the mass of the body (Art. 11), and denote it by M.

We have, from Equation (102),

$$Mkl = \Sigma(mr^2) = K.$$

Substituting for M, l, and k, the values already found, and the value of K will be the moment of inertia, with respect to the assumed axis. Subtract from this the value of Mk^2, and the remainder will be the moment of inertia with respect to a parallel axis through the centre of gravity.

The moment of inertia of a homogeneous body of regular figure, is most readily found by means of the calculus. A few examples of the application of the calculus to finding the moment of inertia of bodies are subjoined.

Application of the Calculus to determine the Moment of Inertia.

131. To render Formula (104) suitable to the application of the calculus, we have simply to change the sign of summation, Σ, to that of integration, $\int$, and to replace m by dM, and r by x. This gives,

$$K = \int x^2 dM \quad \ldots \quad (106.)$$

Example 1. To find the moment of inertia of a rod or bar of uniform thickness with respect to an axis through its centre of gravity and perpendicular to the length of the rod.

Let AB represent the rod, G its centre of gravity, and E any element contained by planes at right angles to the length of the rod and infinitely near each other. Denote the mass of the rod by M, its length, by $2l$, the distance GE, by x, and the thickness of the element E, by dx. Then will the mass of the element E be equal to

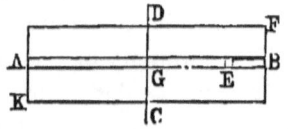

Fig. 110.

MOMENT OF INERTIA.

$\frac{M}{2l}dx$. Substituting this for dM, in Equation (106), and integrating between the limits $-l$ and $+l$, we have,

$$K = \int_{-l}^{+l} \frac{M}{2l} x^2 dx = M\frac{l^2}{3}.$$

For any parallel axis whose distance from G is d, we shall have,

$$K' = M\left(\frac{l^2}{3} + d^2\right) \quad \ldots \quad (107.)$$

These two formulas are entirely independent of the breadth of the filament in the direction of the axis DC. They will, therefore, hold good when the filament AB is replaced by the rectangle KF. In this case, M becomes the mass of the rectangle, $2l$ the length of the rectangle, and d the distance of the centre of gravity of the rectangle from the axis parallel to one of its ends.

Example 2. To find the moment of inertia of a thin circular plate about one of its diameters.

Let ACB represent the plate, AB the axis, and $C''D'$ any element parallel to AB. Denote the radius OC, by r, the distance OE, by x, the breadth of the element EF, by dx, and its length DC, by $2y$. If we denote the entire mass of the plate, by M, the mass of the element CD will be equal to $M\frac{2y\,dx}{\pi r^2}$; or, since $y = \sqrt{r^2 - x^2}$, we have,

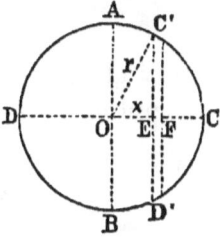

Fig. 111.

$$dM = M\frac{2\sqrt{r^2 - x^2}}{\pi r^2} dx.$$

Substituting in Equation (106), we have,

$$K = \int \frac{M}{\pi r^2} \cdot x^2 (r^2 - x^2)^{\frac{1}{2}} dx.$$

Integrating by the aid of Formulas A and B (Integral Calculus), and taking the integral between the limits $x = -r$, and $x = +r$, we find,

$$K = M \frac{r^2}{4};$$

and for a parallel axis at a distance from AB equal to d,

$$K' = M\left(\frac{r^2}{4} + d^2\right). \quad \ldots \quad (108.)$$

Example 3. To find the moment of inertia of a circular plate with respect to an axis through its centre perpendicular to the face of the plate.

Let the dimensions and mass of the plate be the same as before. Let KL be an elemetary ring whose radius is x, and whose breadth dx. Then will the mass of the elementary ring be equal to $M\frac{2\pi x dx}{\pi r^2}$, or $dM = \frac{2Mx dx}{r^2}$.

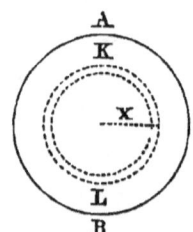

Fig. 112.

Substituting this in Equation (106), and taking the integral between the limits $x = 0$, and $x = r$, we have,

$$K = \int_0^r \frac{2Mx^3 dx}{r^2} = \frac{Mr^2}{2}.$$

For a parallel axis at a distance d from the primitive axis,

$$K' = M\left(\frac{r^2}{2} + d^2\right). \quad \ldots \quad (109.)$$

Example 4. To find the moment of inertia of a circular ring, such as may be generated by revolving a rectangle about a line parallel to one of its sides, taken with respect to an axis through the centre of gravity and perpendicular to the face of the ring. This case differs but little from the preceding. Denote the inner radius by r, the outer radius by r', and the mass of the ring by M. If we take, as before, an elementary ring whose radius is x, and whose breadth is dx, we shall have for its mass,

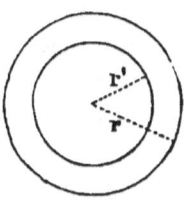

Fig. 113.

$$dM = M \frac{2x \, dx}{r'^2 - r^2}.$$

Substituting in Equation (106), and integrating between the limits r, and r', we have,

$$K = \int_r^{r'} M \frac{2x^3 dx}{r'^2 - r^2} = M \frac{r'^4 - r^4}{2(r'^2 - r^2)} = M \frac{r'^2 + r^2}{2}.$$

For a parallel axis at a distance from the primitive axis equal to d, we have,

$$K' = M \left(\frac{r'^2 + r^2}{2} + d^2 \right) \quad \ldots \quad (110.)$$

If in these values of K and K' we make $r = 0$, we shall deduce the results of the last example.

Example 5. To find the moment of inertia of a right cylinder with respect to an axis through the centre of gravity and perpendicular to the axis of the cylinder.

Let AB represent the cylinder, CD the axis through

its centre of gravity, and E an element of the cylinder between two planes perpendicular to the axis, and distant from each other, by dx. Denote the length of the cylinder by $2l$, the area of its cross section by πr^2, r being the radius of the base; the distance of the section E from the centre of gravity, by x, and the mass of the cylinder, by M.

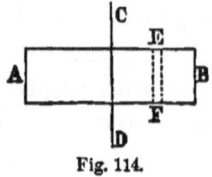

Fig. 114.

The mass of the element E is equal to $M\dfrac{dx}{2l}$. Its moment of inertia with respect to its diameter parallel to CD, is equal to $\dfrac{Mdx}{2l} \times \dfrac{r^2}{4}$ (Example 2), and with respect to CD parallel to it, $\dfrac{Mdx}{2l}\left(\dfrac{r^2}{4} + x^2\right)$.

Integrating this expression between the limits $x = -l$, and $x = +l$, we have,

$$K = \int_{-l}^{+l} \frac{M}{2l}\left(\frac{r^2}{4} + x^2\right) dx = M\left(\frac{r^2}{4} + \frac{l^2}{3}\right).$$

For an axis parallel to the primitive one, and at a distance from it equal to d,

$$K' = M\left(\frac{r^2}{4} + \frac{l^2}{3} + d^2\right) \quad \ldots \ldots \quad (111.)$$

Centre of Gyration.

132. The *centre of gyration* of a body with respect to an axis, is a point at which, if the entire mass be concentrated, its moment of inertia will remain unchanged. The distance from this point to the axis is called the *radius of gyration*.

MOMENT OF INERTIA.

Let M denote the mass of the body, and k' its radius of gyration; then will the moment of inertia of the concentrated mass with respect to the axis, be equal to Mk'^2; but this must, by definition, be equal to the moment of inertia with respect to the same axis, or $\Sigma(mr^2)$; hence,

$$Mk'^2 = \Sigma(mr^2); \quad \text{or,} \quad k' = \sqrt{\frac{\Sigma(mr^2)}{M}} \quad . \quad (112.)$$

That is, *the radius of gyration is equal to the square root of the quotient obtained by dividing the moment of inertia with respect to the same axis, by the entire mass.*

Since M is constant for the same body, it follows that the radius of gyration will be the least possible when the moment of inertia is the least possible, that is, when the axis passes through the centre of gravity. This minimum radius is called the *principal radius of gyration*. If we denote the principal radius of gyration by k, we shall have, from the examples of Article (131), the following results:

Example 1, $\quad k' = \sqrt{\dfrac{l^2}{3} + d^2}; \qquad k = l\sqrt{\tfrac{1}{3}}.$

Example 2, $\quad k' = \sqrt{\dfrac{r^2}{4} + d^2}; \qquad k = \dfrac{r}{2}.$

Example 3, $\quad k' = \sqrt{\dfrac{r^2}{2} + d^2}; \qquad k = r\sqrt{\tfrac{1}{2}}.$

Example 4, $\quad k' = \sqrt{\dfrac{r'^2 + r^2}{2} + d^2}; \quad k = \sqrt{\dfrac{r'^2 + r^2}{2}}$

Example 5, $\quad k' = \sqrt{\dfrac{r^2}{4} + \dfrac{l^2}{3} + d^2}; \quad k = \sqrt{\dfrac{r^2}{4} + \dfrac{l^2}{3}}.$

CHAPTER VI.

CURVILINEAR AND ROTARY MOTION.

Motion of Projectiles.

133. If a body is projected obliquely upwards in vacuum, and then abandoned to the force of gravity, it will be continually deflected from a rectilinear path, and, after describing a curvilinear trajectory, will finally reach the horizontal plane from which it started.

The starting point is called the *point of projection;* the distance from the point of projection to the point at which the projectile again reaches the horizontal plane, through the point of projection, is called the *range*, and the time occupied is called *the time of flight.* The only forces to be considered, are the *initial impulse* and *the force of gravity.* Hence, the trajectory will lie in a vertical plane passing through the line of direction of the initial impulse. Let CAB represent this plane, A the point of projection, AB the range, and AC a vertical line through

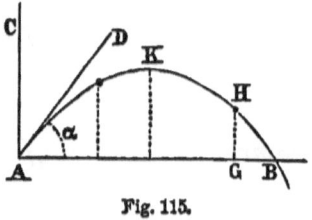

Fig. 115.

A. Take AB and AC as co-ordinate axes; denote the angle of projection DAB, by α, and the velocity due to the initial impulse, by v. Resolve the velocity v into two components, one in the direction AC, and the other in the direction AB. We shall have, for the former, $v \sin\alpha$, and, for the latter, $v \cos\alpha$.

The velocities, and, consequently, the spaces described in the direction of the co-ordinate axes, will (Art. 18) be entirely independent of each other. Denote the space described in the direction AC, in any arbitrary time t, by y. The circumstances of motion in this direction, are those of a body projected vertically upwards with an initial velocity $v \sin\alpha$, and then continually acted upon by the force of gravity. Hence, Equation (78) is applicable. Making, in that equation, $h = y$, and $v' = v \sin\alpha$, we have,

$$y = v \sin\alpha \, t - \tfrac{1}{2} g t^2 \quad . \quad . \quad . \quad . \quad (113.)$$

Denote the space described in the direction of the axis AB, in any arbitrary time t, by x. The only force acting in the direction of this axis, is the component of the initial impulse. Hence, the motion in the direction of the axis of x will be uniform, and Equation (55) is applicable. Making $s = x$, and $v = v \cos\alpha$, we have,

$$x = v \cos\alpha \, t \quad . \quad . \quad . \quad . \quad (114.)$$

If we suppose t to be the same in Equations (113) and (114), they will be simultaneous, and, taken together, will make known the position of the projectile at any instant.

From (114), we have,

$$t = \frac{x}{v \cos\alpha},$$

which, substituted in (113), gives,

$$y = \frac{\sin\alpha}{\cos\alpha} x - \frac{g x^2}{2 v^2 \cos^2\alpha} \quad . \quad . \quad . \quad (115.)$$

an equation which is entirely independent of t. It, therefore, expresses the relation between x and y for any value of t whatever, and is, consequently, the equation of the trajectory. Equation (115) is the equation of a parabola whose axis is vertical. Hence, the required trajectory is a parabola.

To find an expression for the range, make $y = 0$, in (115), and deduce the corresponding value of x. Placing the value of y equal to 0, we have,

$$\frac{\sin\alpha}{\cos\alpha}x - \frac{gx^2}{2v^2\cos^2\alpha} = 0;$$

$$\therefore\ x = 0,\ \text{and}\ x = \frac{2v^2\sin\alpha\cos\alpha}{g}.$$

The first value of x corresponds to the point of projection, and the second is the value of the *range*, AB.

From trigonometry, we have,

$$2\sin\alpha\cos\alpha = \sin 2\alpha.$$

If we denote the height due to the initial velocity, by h, we shall have,

$$v^2 = 2gh.$$

Substituting these in the second value of x, and denoting the range by r, we have,

$$r = 2h\sin 2\alpha\ \ \ .\ \ .\ \ .\ \ .\ \ .\ \ (116.)$$

The greatest value of r will correspond to the value $\alpha = 45°$, in which case, $2\alpha = 90°$, and $\sin 2\alpha = 1$. Hence, we have, for the greatest range,

$$r = 2h.$$

That is, it is equal to *twice the height due to the initial velocity*.

If, in (116), we replace α by $90° - \alpha$, we shall have,

$$r = 2h\sin(180° - 2\alpha) = 2h\sin 2\alpha,$$

the same value as before. Hence, we conclude that there are two angles of projection, complements of each other, which give the same range. The trajectories in the two cases are not the same, as may be shown by substituting the values of α, and 90° — α, in Equation (115). The greater angle of projection gives a higher elevation, and, consequently, the projectile descends more vertically. It is for this reason that the gunner selects the greater of the two angles of elevation when he desires to crush an object, and the lesser one when he desires to batter, or overturn the object. If α = 90°, the value of r becomes 0. That is, if a body be projected vertically upwards, it will return to the point of projection.

To find *the time of flight*, make $x = r$, in Equation (114), and deduce the corresponding value of t. This gives,

$$t = \frac{r}{v \cos\alpha} \quad \ldots \quad (117.)$$

The range being the same, the time of flight will be greatest when α is greatest. Equation (114) also gives the time required for the body to describe any distance in the direction of the horizontal line AB.

In Equation (117) there are four quantities, t, r, v, and α, and from it, if any three are given, the remaining one may be determined.

As an application of the principles just deduced, let it be required to determine the angle of projection, in order that the projectile may strike a point H, at a horizontal distance $AG = x'$ from the point of projection, and at a height $GH = y'$ above it.

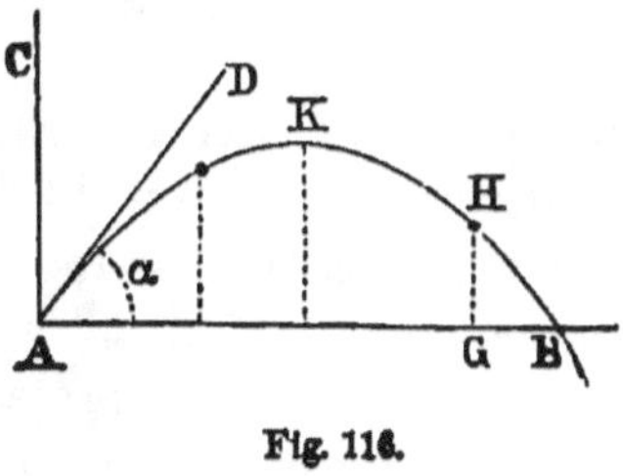

Since the point H lies on the trajectory, its co-ordinates must satisfy the equation of the curve, giving

Fig. 116.

$$y' = x' \tan\alpha - \frac{gx'^2}{2v^2 \cos^2\alpha}.$$

From trigonometry, we have,

$$\cos^2\alpha = \frac{1}{\sec^2\alpha} = \frac{1}{1+\tan^2\alpha}.$$

Substituting this in the preceding equation, we have, after clearing of fractions,

$$2v^2 y' = 2v^2 x' \tan\alpha - gx'^2(1+\tan^2\alpha);$$

or, transposing and reducing,

$$\tan^2\alpha - \frac{2v^2}{gx'}\tan\alpha = -\frac{2v^2 y' + gx'^2}{gx'^2}.$$

Hence,

$$\tan\alpha = \frac{v^2}{gx'} \pm \sqrt{\frac{v^4}{g^2 x'^2} - \frac{2v^2 y' + gx'^2}{gx'^2}};$$

or, making $v^2 = 2gh$,

$$\tan\alpha = \frac{2h}{x'} \pm \sqrt{\frac{4h^2}{x'^2} - \frac{4hy' + x'^2}{x'^2}} = \frac{2h \pm \sqrt{4h^2 - 4hy' - x'^2}}{x'}.$$

This shows that there are, in general, two angles of projection, under either of which the point may be struck.

If we suppose

$$x'^2 = 4h^2 - 4hy' \quad \ldots \quad (118.)$$

the quantity under the radical sign will be 0, and the two angles of projection will become one.

But if x' and y' be regarded as variables, Equation (118) represents a parabola whose axis is a vertical passing through the point of projection. Its vertex is at a distance above the point A, equal to h, its focus is at A, and its parameter is equal to $4h$, or twice the range.

If we suppose

$$x'^2 < 4h^2 - 4hy',$$

the point (x', y'), will lie within the parabola just described, the quantity under the radical sign will be positive, and there will be two real values of $\tan \alpha$, and, consequently, two angles of projection, under either of which the point may be struck.

If we suppose

$$x'^2 > 4h^2 - 4hy',$$

the point (x', y'), will be without this parabola, the values of $\tan \alpha$ will both be imaginary, and there will be no angle under which the point can be struck.

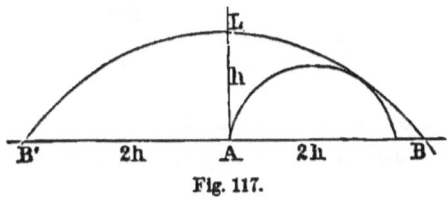

Fig. 117.

Let the parabola $B'LB$ represent the curve whose equation is

$$x'^2 = 4h^2 - 4hy'.$$

Conceive it to be revolved about AL, as an axis generating a paraboloid of revolution. Then, from what has preceded, we conclude, first, that every point lying within the surface may be reached from A, with a given initial velocity, under two different angles of projection; second, that every point lying on the surface can be reached, but only by a single angle of projection; thirdly, that no point lying without the surface can be reached at all.

If we suppose a body to be projected horizontally from an elevated point A, the trajectory will be made known by Equation (115) by simply making $\alpha = 0$; whence, $\sin\alpha = 0$, and $\cos\alpha = 1$. Substituting and reducing, we have,

$$y = -\frac{gx^2}{2v^2} \quad . \quad (119.)$$

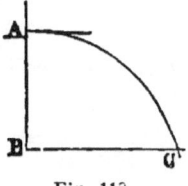

Fig. 118.

For every value of x, y is negative, which shows that every point of the trajectory lies below the horizontal line through the point of projection. If we suppose ordinates to be estimated positively downwards, we shall have,

$$y = \frac{gx^2}{2v^2} \quad \ldots \quad (120.)$$

To find the point at which the trajectory will reach any horizontal plane BC, whose distance below the point A is h', we make $y = h'$, in (120), whence,

$$x = BC = v\sqrt{\frac{2h'}{g}} \quad \ldots \quad (121.)$$

On account of the resistance of the air, the results of the preceding discussion will be greatly modified. They will, however, approach more nearly to the observed phenomena, as the velocity is diminished and the density of the projectile increased. The atmospheric resistance increases as the square of the velocity, and as the cross section of the projectile exposed to the action of the resistance. In the air, it is found that, under ordinary circumstances, the maximum range is obtained by an angle of projection not far from 34°.

EXAMPLES.

1. What is the time of flight of a projectile, when the angle of projection is 45°, and the range 6000 feet?

SOLUTION.

When the angle of projection is 45°, the range is equal to twice the height due to the velocity of projection. Denoting this velocity by v, we shall have,

$$v^2 = 2gh = 2 \times 32\tfrac{1}{6} \times 3000 = 193000.$$

CURVILINEAR AND ROTARY MOTION.   195

Whence, we find,
$$v = 439.3 \text{ ft.}$$

From Equation (117), we have,

$$t = \frac{r}{v \cos \alpha} = \frac{6000}{439.3 \cos 45°} = 19.3 \text{ sec. } Ans.$$

2. What is the range of a projectile, when the angle of projection is 30°, and the initial velocity 200 feet?
Ans. 1076.9 ft.

3. The angle of projection under which a shell is thrown is 32°, and the range 3250 feet. What is the time of flight?
Ans. 11.25 sec., nearly.

4. Find the angle of projection and velocity of projection of a shell, so that its trajectory shall pass through two points, the co-ordinates of the first being $x = 1700$ ft., $y = 10$ ft., and of the second, $x = 1800$ ft., $y = 10$ ft.

SOLUTION.

Substituting for x and y, in Equation (115), (1700, 10), and (1800, 10), we have,

$$10 = 1700 \tan \alpha - \frac{(1700)^2 g}{2v^2 \cos^2 \alpha};$$

and,

$$10 = 1800 \tan \alpha - \frac{(1800)^2 g}{2v^2 \cos^2 \alpha}.$$

Finding the value of $\frac{g}{2v^2 \cos^2 \alpha}$ from each of these equations, and placing the two equal to each other, we have, after reduction,

$$(18)^2(1 - 170 \tan \alpha) = (17)^2(1 - 180 \tan \alpha).$$

196   MECHANICS.

Whence, by solution,

$$\tan\alpha = \tfrac{7}{612} = 0.01144, \text{ nearly}; \quad \therefore \; \alpha = 39'\,19''.$$

We have, from trigonometry,

$$\cos^2\alpha = \frac{1}{\sec^2\alpha} = \frac{1}{1+\tan^2\alpha} = \frac{374544}{374593} = .99987.$$

Substituting for $\tan\alpha$ and $\cos\alpha$ in the first equation their values as just deduced, we find, for v^2,

$$v^2 = \frac{(1700)^2 g}{2\cos^2\alpha(1700\tan\alpha - 10)} = \frac{92961666}{18.89} = 4921210.$$

Whence,
$$v = 2218.3 \text{ ft.}$$

The required angle of projection is, therefore, $39'19''$, and the required initial velocity, 2218.3 ft.

4. At what elevation must a shell be projected with a velocity of 400 feet, that it may range 7500 feet on a plane which descends at an angle of 30?

SOLUTION.

The co-ordinates of the point at which the shell strikes, are

$$x' = 7500\cos 30° = 6495; \text{ and } y' = -7500\sin 30° = -3750.$$

And denoting the height due to the velocity 400 ft., by h, we have,

$$h = \frac{v^2}{2g} = 2486 \text{ ft.}$$

Substituting these values in the formula,

$$\tan\alpha = \frac{2h \pm \sqrt{4h^2 - 4hy' - x'^2}}{x'},$$

and reducing, we have,

$$\tan\alpha = \frac{4972 \pm 4453}{6495}.$$

Hence, $\alpha = 4°\ 34'\ 10''$, and $55°\ 25'\ 41''$. *Ans.*

Centripetal and Centrifugal Forces.

134. Curvilinear motion can only result from the action of an incessant force, whose direction differs from that of the original impulse. This force is called the *deflecting force*, and may arise from one or more active forces, or it may result from the resistance offered by a rigid body, as when a ball is compelled to run in a curved groove. Whatever may be the nature of the deflecting forces, we can always conceive them to be replaced by a single incessant force acting transversely to the path of the body. Let the deflecting force be resolved into two components, one normal to the path of the body, and the other tangential to it. The latter force will act to accelerate or retard the motion of the body, according to the direction of the deflecting force; the former alone is effective in changing the direction of the motion. The normal component is always directed towards the concave side of the curve, and is called the *centripetal force*. The body resists this force, by virtue of its inertia, and, from the law of inertia, the resistance must be *equal and directly opposed* to the centripetal force. This force of resistance is called *the centrifugal force*. Hence, we may define the centrifugal force to be *the resistance which a body offers to a force which tends to deflect it from a rectilineal path*. The centripetal and centrifugal forces taken together, are called central forces.

Measure of the Centrifugal Force.

135. To deduce an expression for the measure of the centrifugal force, let us first consider the case of a single material point, which is constrained to move in a circular path by a force constantly directed towards the centre, as when a solid body is confined by a string and whirled around a fixed point. In this case, the tangential component of the deflecting force is always 0. There will be no loss of velocity in consequence of a change of direction in the motion (Art. 120). Hence, the motion of the point will be uniform.

Let ABD represent the path of the body, and V its centre. Suppose the circumference of the circle to be a regular polygon, having an infinite number of sides, of which AB is one; and denote each of these sides by ds. When the body reaches A, it tends, by virtue of its inertia, to move in the direction of the tangent AT; but, in consequence of the action of the centripetal force directed towards V, it is constrained to describe the side ds in the time dt. If we draw BC parallel to AT, it will be perpendicular to the diameter AD, and AC will represent the space through which the body has been drawn from the tangent, in the time dt. If we denote the acceleration due to the centripetal force by f, and suppose it to be constant during the time dt, we shall have, from Art. 114,

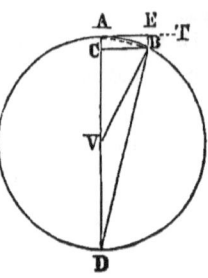

Fig. 119.

$$AC = \tfrac{1}{2}fdt^2 \quad \ldots \ldots \quad (122.)$$

From a property of right-angled triangles, we have, since $AB = ds$,

$$ds^2 = AC \times AD; \quad \text{or,} \quad ds^2 = AC \times 2r.$$

Whence,

$$AC = \frac{ds^2}{2r}.$$

Substituting this value of AC in (122), and solving with respect to f,

$$f = \frac{ds^2}{dt^2} \times \frac{1}{r}.$$

But $\frac{ds^2}{dt^2} = v^2$ (Art. 113), in which v denotes the velocity of the moving point. Substituting in the preceding equation, we have,

$$f = \frac{v^2}{r} \quad \ldots \quad (123.)$$

Here f is the acceleration due to the deflecting force; and, since this is exactly equal to the centrifugal force, we have *the acceleration due to the centrifugal force equal to the square of the velocity, divided by the radius of the circle.*

If the mass of the body be denoted by M, and the *entire centrifugal* force by F, we shall have (Art. 24),

$$F = \frac{Mv^2}{r} \quad \ldots \quad (124.)$$

If we suppose the body to be moving on any curve whatever, we may, whilst it is passing over any two consecutive elements, regard it as moving on the arc of the osculatory circle to the curve which contains these elements; and, further, we may regard the velocity as uniform during the infinitely small time required to describe these elements. The direction of the centrifugal force being normal to the curve, must pass through the centre of the osculatory circle. Hence, all the circumstances of motion are the same as before, and Equations (123) and (124) will be applicable, provided r be taken as the radius of the curvature. Hence, we may enunciate the law of the centrifugal force as follows:

The acceleration due to the centrifugal force is equal to the square of the velocity of the body divided by the radius of curvature.

The entire centrifugal force is equal to the acceleration, multiplied by the mass of the body.

In the case of a body whirled around a centre, and restrained by a string, the tension of the string, or the force exerted to break it, will be measured by the centrifugal force. The radius remaining constant, the tension will increase as the square of the velocity.

Centrifugal Force at points of the Earth's Surface.

136. Let it be required to determine the centrifugal force at different points of the earth's surface, due to its rotation on its axis.

Suppose the earth spherical. Let A be any point on the surface, PQP' a meridian section through A, PP' the axis, FQ the equator, and AB perpendicular to PP', the radius of the parallel of latitude through A. Denote the radius of the earth by r, the radius of the parallel through A by r', and the latitude of A, or the angle ACQ, by l. The time of

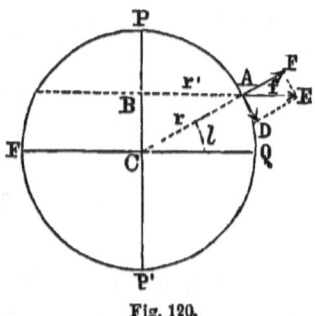

Fig. 120.

revolution being the same for every point on the earth's surface, the velocities of Q and A will be to each other as their distances from the axis. Denoting these velocities by v and v', we have,

$$v : v' :: r : r',$$

whence,

$$v' = \frac{vr'}{r}.$$

But, from the right-angled triangle CAB, since the angle at A is equal to l, we have,

$$r' = r \cos l.$$

Substituting this value of r' in the value of v', and reducing, we have,

$$v' = v \cos l.$$

If we denote the acceleration due to the centrifugal force at the equator by f, we shall have, Equation (123),

$$f = \frac{v^2}{r} \quad \ldots \ldots \quad (125.)$$

In like manner, if we denote the acceleration due to the centrifugal force at A, by f', we shall have,

$$f' = \frac{v'^2}{r'}.$$

Substituting for v' and r' their values, previously deduced, we get,

$$f' = \frac{v^2 \cos l}{r} \quad \ldots \ldots \quad (126.)$$

Comparing Equations (125) and (126), we find,

$$f : f' :: 1 : \cos l, \quad \therefore \; f' = f \cos l \quad (127.)$$

That is, *the centrifugal force at any point on the earth's surface is equal to the centrifugal force at the equator, multiplied by the cosine of the latitude of the place.*

Let AE, perpendicular to PP', represent the value of f', and resolve it into two components, one tangential, and the other normal to the meridian section. Prolong CA, and draw AD perpendicular to it at A. Complete the rectangle FD on AE as a diagonal. Then will AD represent the tangential, and AF the normal component of f'. In the right-angled triangle AFE, the angle at A is equal to l. Hence,

$$FE = AD = f' \sin l = f \cos l \sin l = \frac{f \sin 2l}{2} \quad (128.)$$

$$AF = f' \cos l = f \cos^2 l \quad \ldots \ldots \quad (129.)$$

From (128), we conclude that the tangential component is

0 at the equator, goes on increasing till $l = 45°$, where it is a maximum; then goes on decreasing till the latitude is 90° when it again becomes 0.

The effect of the tangential component is to heap up the particles of the earth about the equator, and, were the earth in a fluid state, this process would go on till the effect of the tangential component was exactly counterbalanced by component of gravity acting down the inclined plane thus found, when the particles would be in a state of equilibrium. The higher analysis has shown that the form of equilibrium is that of an oblate spheroid, differing but slightly from that which our globe is found to possess by actual measurement.

From Equation (129), we see that the normal component of the centrifugal force is equal to the centrifugal force at the equator multiplied by the square of the cosine of the latitude of the place.

This component is directly opposed to gravity, and, consequently, tends to diminish the weight of all bodies on the surface of the earth. The value of this component is greatest at the equator, and diminishes towards the poles, where it becomes equal to 0. From the action of the normal component of the centrifugal force, and from the flattened form of the earth due to the tangential component bringing the polar regions nearer the centre of the earth, the measured force of gravity ought to increase in passing from the equator towards the poles. This is found, by observation, to be the case.

The radius of the earth at the equator is found, by measurement, to be about 3962.8 miles, which, multiplied by 2π, will give the entire circumference of the equator. If this be divided by the number of seconds in a day, 86400, we find the value of v. Substituting this value of v and that of r just given, in Equation (125), we should find,

$$f = 0.1112 \text{ ft.,}$$

for the measure of the centrifugal force at the equator. If

this be multiplied by the square of the cosine of the latitude of any place, we shall have the value of the normal component of the centrifugal force at that place.

Centrifugal Force of Extended Masses.

136. We have supposed, in what precedes, the dimensions of the body under consideration to be extremely small; let us next examine the case of a body, of any dimensions whatever, constrained to revolve about a fixed axis, with which it is invariably connected. If we suppose this body to be divided into infinitely small elements, whose directions are parallel to the axis, the centrifugal force of each element will, from what has preceded, be equal to the mass of the element into the square of its velocity, divided by its distance from the axis. If a plane be passed through the centre of gravity of the body, perpendicular to the axis, we may, without impairing the generality of the result, suppose the mass of each element to be concentrated at the point in which this plane cuts the line of direction of the element.

Let XCY be the plane through the centre of gravity of the body perpendicular to the axis of revolution, AB the section cut out of the body, or the projection of the body on the plane, and C the point in which it cuts the axis. Take C as the origin of a system of rectangular axes, and let CX be the axis of X, CY the axis of Y, and let m be the point at which the mass of one of these filaments is concentrated, and denote that mass by m. Denote the co-ordinates of m by x and y, its distance from C by r, and its velocity by v. The centrifugal force of the mass m will be equal to

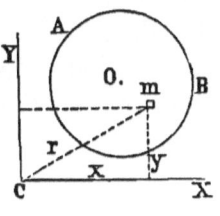

Fig. 121

$$\frac{mv^2}{r}.$$

If we denote the angular velocity of the body by V, the

velocity of the point m will be equal to rV', which, being substituted in the expression for the centrifugal force just deduced, gives

$$mrV'^2.$$

Let this force be resolved into two components, respectively parallel to the axes CX and CY. We shall have, for these components, the expressions,

$$mrV'^2\cos mCX, \quad \text{and} \quad mrV'^2\sin mCX.$$

But from the figure, we have,

$$\cos mCX = \frac{x}{r}, \quad \text{and} \quad \sin mCX = \frac{y}{r}.$$

Substituting these values in the preceding expressions, and reducing, we have, for the two components,

$$mxV'^2, \quad \text{and} \quad myV'^2.$$

In like manner, if we denote the masses of the remaining filaments by m', m'', &c., the co-ordinates of the points at which they are cut by the plane XCY, by x', y'; x'', y'', &c., their distances from the axis by r', r'', &c., and resolve the centrifugal forces into components, respectively parallel to the axes, we shall have, since V' remains the same,

$$m'x'V'^2, \quad m'y'V'^2;$$
$$m''x''V'^2, \quad m''y''V'^2;$$
$$\&c., \quad \&c.$$

If we denote the sum of the components in the direction of the axis of X by X, and in the direction of the axis of Y by Y, we shall have,

$$X = \Sigma(mx)V'^2, \quad \text{and} \quad Y = \Sigma(my)V'^2.$$

If, now, we denote the entire mass of the body, by M, and suppose it concentrated at its centre of gravity O, whose co-ordinates are designated by x_1, and y_1, and whose distance from C is equal to r_1, we shall have, from the principle of the centre of gravity (Art. 51),

$$\Sigma(mx) = Mx_1, \quad \text{and} \quad \Sigma(my) = My_1.$$

Substituting above, we have,

$$X = MV'^2 x_1, \quad \text{and} \quad Y = MV'^2 y_1.$$

If we denote the resultant of all the centrifugal forces, which will be the centrifugal force of the body, by R, we shall have,

$$R = \sqrt{X^2 + Y^2} = MV'^2\sqrt{x_1^2 + y_1^2} = MV'^2 r_1.$$

But if the velocity of the centre of gravity be denoted by V, we shall have,

$$V = V'r_1; \quad \text{or,} \quad V'^2 = \frac{V^2}{r_1^2};$$

which, substituted in the preceding result, gives, for the resultant,

$$R = \frac{MV^2}{r_1} \quad \ldots \ldots \quad (130.)$$

The line of direction of R is made known by the equations,

$$\cos a = \frac{X}{r_1}, \quad \text{and} \quad \cos b = \frac{Y}{r_1};$$

it, therefore, passes through the centre of gravity O.

Hence, we conclude, *that the centrifugal force of an extended mass, constrained to revolve about a fixed axis, with which it is invariably connected, is the same as though the entire mass were concentrated at its centre of gravity.*

Pressure on the Axis.

137. The centrifugal force, passing through the centre of gravity and intersecting the axis, will exert its entire effect in creating a pressure upon the axis of revolution. By inspecting the equation,

$$R = MV'^2 r_1,$$

we see that this pressure will increase with the mass, the angular velocity, and the distance of the centre of gravity from the axis. When the last distance is 0, that is, when the axis of revolution passes through the centre of gravity, there will be no pressure on the axis arising from the centrifugal force, no matter what may be the mass of the body or its angular velocity. Such is the case of the earth revolving on its axis.

Principal Axes.

138. Suppose the axis about which a body revolves to become free, so that the body can move in any direction. If that axis be not one of symmetry, it will be pressed unequally in different directions by the centrifugal force, and will immediately alter its position. The body will for an instant rotate about some other line, which will immediately change its position, giving place to a new axis of rotation, which will instantly change its position, and so on, until an axis is reached which is pressed equally in all directions by the centrifugal forces of the elements. The body will then continue to revolve about this line, by virtue of its inertia, until the revolution is destroyed by the action of some extraneous force. Such an axis is called a *principal axis of rotation*. Every body has at least *one* such axis, and may have more. The axis of a cone or cylinder is a *principal axis;* any diameter of a sphere is a *principal axis;* in short, any axis of symmetry of a homogeneous solid is a *principal axis*. The shortest axis of an oblate spheroid is a principal axis; and it is found by observation that all of the planets of the solar system, which are oblate spheroids, revolve about their shorter axes, whatever may be the inclination of these axes to the planes of their orbits. Were the earth, by the action of any extraneous force, constrained to revolve about some other axis than that about which it is found to revolve, it would, as soon as the force ceased to act, return to its present axis of rotation.

Experimental Illustrations.

139. The principles relating to the centrifugal force admit of experimental illustration. The instrument represented in the figure, may be employed to show the value of the centrifugal force. A represents a vertical axle upon which is mounted a wheel F, communicating with a train of wheel-work, by means of which the axle may be made to revolve with any angular velocity. At the upper end of the axle is a forked branch BC, sustaining a stretched wire. D and E are two balls which are pierced by the wire, and are free to move along it. Between B and E is a spiral spring, whose axis coincides with the wire.

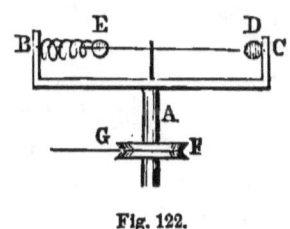

Fig. 122.

Immediately below the spring, on the horizontal part of the fork, is a scale for determining the distance of the ball E, from the axis, and for measuring the degree of compression of the spring. Before using the instrument, the force required to produce any degree of compression of the spring is determined experimentally, and marked on the scale.

If now a motion of rotation be communicated to the axis, the ball D will at once recede to C, but the ball E will be restrained by the spiral spring. As the velocity of rotation is increased, the spring will be compressed more and more, and the ball E, will approach B. By a suitable arrangement of the wheelwork, the angular velocity of the axis corresponding to any degree of compression may be ascertained. We have thus all the data necessary to a verification of the law of the centrifugal force.

If a vessel of water be made to revolve about a vertical axis, the interior particles will recede from the axis on account of the centrifugal force, and will be heaped up about the sides of the vessel, imparting a concave form to the upper surface. The concavity will become greater as the angular velocity is increased.

If a circular hoop of flexible metal be fastened so that one of its diameters shall coincide with the axis of a whirling machine, its lower point being fastened to the horizontal beam, and a motion of rotation be imparted, the portions of the hoop farthest from the axis will be most affected by the centrifugal force, and the hoop will be observed to assume an elliptical form.

If a sponge, filled with water, be attached to one of the arms of a whirling machine, and a motion of rotation be imparted, the water will be thrown from the sponge. This principle has been made use of in a machine for drying clothes. An annular trough of copper is mounted upon an axis by means of radial arms, the axis being connected with a train of wheelwork, by means of which it may be put in motion. The outer wall is pierced with holes for the escape of the water, and a lid serves to confine the articles to be dried. To use this instrument, the linen, after being washed, is placed in the annular space, and a rapid motion of rotation imparted to the machine. The linen is thrown, by the centrifugal force, against the outer wall of the instrument, and the water, being partially squeezed out, and partially thrown off by the centrifugal force, escapes through the holes made for the purpose. Sometimes as many as 1,500 revolutions per minute are given to the drying machine, in which case, the drying process is very rapid and very perfect.

If a body be whirled about an axis with sufficient velocity, it may happen that the centrifugal force generated will be greater than the force of cohesion which binds the particles together, in which case, the body will be torn asunder. It is a common occurrence that large grindstones, when put into a state of rapid rotation, burst, the fragments being thrown with great velocity away from the axis, and often producing much destruction.

When a wagon, or carriage, is driven rapidly around a corner, or is forced to turn about a circular track, the centrifugal force generated is often sufficient to throw out the loose articles from the vehicle, and even to overthrow the vehicle itself. When a car upon a railroad track is forced to turn around a sharp curve, the centrifugal force generated, tends to throw the weight of the cars against the rail, producing a great amount of friction, and contributing to wear out both the track and the car. To obviate this difficulty in a measure, it is customary to raise the outer rail, so that the resultant of the centrifugal force, and the force of gravity, shall be sensibly perpendicular to the plane of the two rails.

Elevation of the outer rail of a curved track.

140. To find the inclination of the track, that is, the elevation of the outer rail, so that the resultant of the weight and centrifugal force may be perpendicular to the line joining the two rails. Let G be the centre of gravity of the car, and let the figure represent a vertical section through the centre of gravity and the centre of the curved track. Let GA, parallel to the horizon, represent the ac-

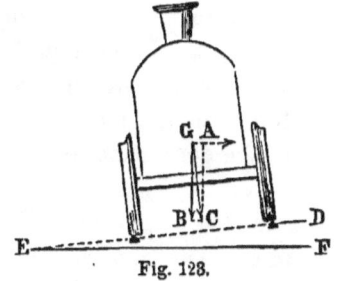

Fig. 128.

celeration due to the centrifugal force, and GB, perpendicular to the horizon, the acceleration due to the weight of the car. Construct the resultant GC, of these forces, then must the line DE be perpendicular to GC. Denote the velocity of the car, by v, and the radius of the curved track, by r. The acceleration due to the weight will be equal to g, the force of gravity, and the acceleration due to the centrifugal force will be equal to $\frac{v^2}{r}$. The tangent of the angle CGB will be equal to $\frac{CB}{GB}$; or, denoting the angle by α, we shall have,

$$\tan\alpha = \frac{CB}{GB} = \frac{v^2}{gr}.$$

But the angle DEF is equal to the angle CGB. Denoting the distance between the rails, by d, and the elevation of the outer rail above the inner one, by h, we shall have,

$$\tan\alpha = \frac{h}{d}, \text{ very nearly.}$$

Equating the two values of $\tan\alpha$, we have,

$$\frac{h}{d} = \frac{v^2}{gr}, \quad \therefore h = \frac{dv^2}{gr} \quad . \quad (131.)$$

Hence, the elevation of the outer rail varies as the square of the velocity directly, and as the radius of the curve inversely.

It is obvious that this correction would require to be different for different velocities, which, from the nature of the case, would be manifestly impossible. The correction is, therefore, made for some assumed velocity, and then such a form is given to the tire of the wheels as will complete the correction for different velocities.

The Conical Pendulum.

141. The conical pendulum consists of a solid ball attached to one end of a rod, the other end of which is connected, by means of a hinge-joint, with a vertical axle. When the axle is put in motion, the centrifugal force generated in the ball causes it to recede from the axis, until an equilibrium is established between the weight of the ball, the centrifugal force, and the tension of the connecting rod.

When the velocity is constant, the centrifugal force will be constant, and the centre of the ball will describe a horizontal circle, whose radius will depend upon the velocity. Let it be required to determine the time of revolution.

Let BD be the vertical axis, A the ball, B the hinge-joint, and AB the connecting rod, whose mass is so small, that it may be neglected, in comparison with that of the ball.

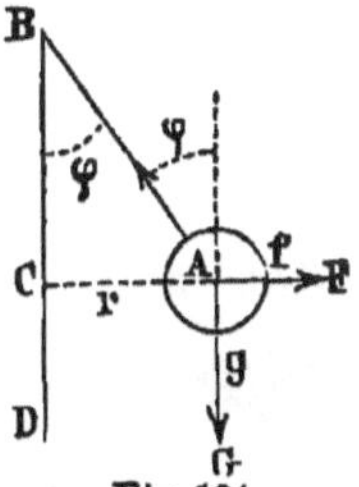

Denote the required time of revolution, by t, the length of the arm, by l, the acceleration due to the centrifugal force, by f, and the angle ABC, by φ. Draw AC perpendicular to BD, and denote AC, by r, and BC, by h.

Fig. 124.

From the triangle ABC, we have, $r = l\sin\varphi$; and since r is the radius of the circle described by A, we have the distance passed over by A, in the time t, equal to $2\pi r = 2\pi l\sin\varphi$. Denoting the velocity of A, by v, we have, from Equation (55),

$$v = \frac{2\pi l \sin\varphi}{t}.$$

But the centrifugal force is equal to the square of the velocity, divided by the radius; hence,

$$f = \frac{4\pi^2 l \sin\varphi}{t^2} \quad \ldots \quad (132.)$$

The forces which act upon A, are the centrifugal force in the direction AF, the force of gravity in the direction AG, and the tension of the connecting rod in the direction AB. In order that the ball may remain at an invariable distance from the axis, these three forces must be in equilibrium. Hence (Art. 35),

$$g : f : : \sin BAF : : \sin BAG ;$$

but, $\qquad \sin BAF = \sin(90° + \varphi) = \cos\varphi$;

and, $\quad \sin BAG = \sin(180° - \varphi) = \sin\varphi$;

whence, by substitution,

$$g : f :: \cos\varphi : \sin\varphi, \quad \therefore g = f\frac{\cos\varphi}{\sin\varphi}.$$

Substituting for f its value, taken from (132), we have,

$$g = \frac{4\pi^2 l\cos\varphi}{t^2}.$$

But, from the triangle ABC, we have, $l\cos\varphi = h$, which gives,

$$g = \frac{4\pi^2 h}{t^2}, \quad \therefore t = 2\pi\sqrt{\frac{h}{g}} \quad . \quad . \quad (133.)$$

That is, the time of a revolution is equal to the time of a double vibration of a pendulum whose length is h.

The Governor.

142. The principle of the conical pendulum is employed in the *governor*, a machine attached to engines, to regulate the motive force.

AB is a vertical axis connected with the machine near its working point, and revolving with a velocity proportional to that of the working point; FE and GD are two arms turning freely about AB, and bearing heavy balls D and E, at their extremities; these bars are united by hinge-joints with two other bars at G and F, these bars are also attached to a ring at H, free to slide up and down the shaft.

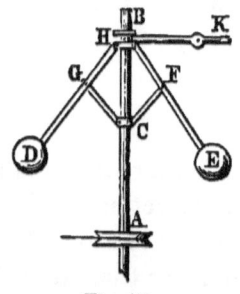

Fig. 125.

The governor is so constructed, that the figure $GCFH$ is always a parallelogram. The ring at H is connected with a lever HK, which may be made to act upon the valve that admits steam to the cylinder.

When the shaft revolves, the centrifugal force developed in the balls, causes them to recede from the axis, and the ring H is depressed; and when the velocity has become sufficiently great, the lever begins to act in closing the valve. If the velocity slackens, the balls approach the axis, and the ring H ascends, opening the valve again. In any given case, if we know the velocity required at the working point, we can from it compute the required angular velocity of the shaft, and, consequently, the value of t. This value of t being substituted in Equation (133), makes known the value of h. We may, therefore, make the proper adaptation of the ring, and of the lever HK.

EXAMPLES.

1. A ball weighing 10 lbs. is whirled around in a circle whose radius is 10 feet, with a velocity of 30 feet per second. What is the acceleration of the centrifugal force?

Ans. 90 ft.

2. In the preceding example, what is the tension upon the cord which restrains the ball?

SOLUTION.

Denote the tension in pounds, by t; then, since the pressures produced by two forces are proportional to their accelerations, we shall have,

$10 : t :: g : 90,$ $\therefore t = 28$ lbs., nearly. *Ans.*

3. A body is whirled around in a circular path whose radius is 5 feet, and it is observed that the pressure due to the centrifugal force is just equal to the weight of the body. What is the velocity of the moving body?

SOLUTION.

Denoting the velocity by v, we have the acceleration due to the centrifugal force equal to $\frac{v^2}{5}$; but, by the conditions of the problem, this is equal to the acceleration due to the weight of the body. Hence,

$$\frac{v^2}{5} = g = 32\tfrac{1}{6}, \quad \therefore v = 12.7 \text{ ft.} \quad Ans.$$

4. In how many seconds must the earth revolve on its axis in order that the centrifugal force at the equator may exactly counterbalance the force of gravity, the radius of the equator being taken equal to 3962.8 miles?

SOLUTION.

Reducing the miles to feet, and denoting the required velocity, by v, we have,

$$\frac{v^2}{20923584} = 32\tfrac{1}{6}, \quad \therefore v = \sqrt{32\tfrac{1}{6} \times 20923584}.$$

But the time of revolution is equal to the circumference of the equator, divided by the velocity. Denoting the time by t, we have,

$$t = \frac{2\pi \times 20923584}{v};$$

and, substituting the value of v, taken from the preceding equation, we have, after reduction,

$$t = \frac{2\pi \sqrt{20923584}}{\sqrt{32\tfrac{1}{6}}} = \frac{2\pi \times 4574}{5.67} = 5068 \text{ secs.} \quad Ans.$$

But the earth actually revolves in 86400 sideral, or in about 86164 mean solar seconds. Hence, the earth would have to revolve 17 times as fast as at present, in order that the centrifugal force at the equator might be equal to the force of gravity.

5. A body is placed on a horizontal plane, which is made to revolve about a vertical axis, with an angular velocity of 2 feet. How far must the body be situated from the axis that it may be on the point of sliding outwards, the coefficient of friction between the body and plane being equal to .6?

SOLUTION.

Denote the required distance by r; then will the velocity of the body be equal to $2r$, and the acceleration due to the centrifugal force will be equal to $4r$. But the acceleration due to the force of friction is equal to $0.6 \times g = 19.3$ ft. From the conditions of the problem, these two are equal, hence,

$$4r = 19.3 \text{ ft.}, \quad \therefore \ r = 4.825 \text{ ft.} \quad Ans.$$

6. What must be the elevation of the outer rail of a railroad track, the radius of curvature being 3960 ft., the distance between the rails 5 feet, and the velocity of the car 30 miles per hour, in order that the centrifugal force may be exactly counterbalanced by the component of the weight parallel to the line joining the rails?

Ans. 0.076 ft., or 0.9 in., nearly.

7. The distance between the rails is 5 feet, the radius of the curve 600 feet, and the height of the centre of gravity of the car 5 feet. What velocity must be given to the car that it may be on the point of being overturned by the centrifugal force, the rails being on the same level?

We have,

$$v = \sqrt{\frac{5 \times 32\tfrac{1}{6} \times 600}{2 \times 5}} = 98 \text{ ft., or } 66\tfrac{3}{4} \text{ m., per hour.} \quad Ans.$$

Work.

143. By the term *work*, in mechanics, is meant the effect produced by a force in overcoming a resistance, such as weight, inertia, &c. The idea of work implies that a force is continually exerted, and that the point at which it is applied moves through a certain space. Thus, when a weight is raised through a vertical height, the power which overcomes the resistance offered by the weight is said to work, and the amount of work performed evidently depends, *first*, upon the weight raised, and, *secondly*, upon the height through which it is raised. All kinds of work may be assimilated to the raising of a weight. Hence it is, that this kind of work is assumed as a standard to which all other kinds of work are referred.

The unit of work most generally adopted in this country, is the effort required to raise one pound through a height of one foot. The number of units of work required to raise any weight to any height will, therefore, be equal to the product obtained by multiplying the number of pounds in the weight by the number of feet in the height. If we take the weight of the body as it would be at the equator, for the sake of uniformity in notation, we may regard the *weight* and the *mass* as identical (Art. 11). If we denote the quantity of work expended in raising a body, by Q, the mass of the body, by m, and the height, by h, we shall have,

$$Q = mh.$$

When very large quantities of work are to be estimated, as in the case of steam-engines and other powerful machines, a different unit is sometimes employed, called a *horse power*. When this unit is employed, time enters as an element. A *horse power* is a power which is capable of raising 33,000 lbs. through a height of one foot in one minute; that is, it is a power capable of performing 33,000 units of work in a minute of time, or 550 units of work in one *second*. When an engine, then, is spoken of as being of 100 horse power, it is to be understood that it is capable of performing 55,000 units of work in a second.

In general, if a force acts to overcome a resistance of m pounds, through a distance of n feet, whatever may be the cause of the resistance, or whatever may be the direction of the motion, the quantity of work will be measured by a unit of work taken mn times.

If the pressure exerted by the force is variable, we may conceive the path described by the point of application to be divided into equal parts, so small that, for each part, the pressure may be regarded as constant. If we denote the length of one of these equal parts, by p, and the force exerted whilst describing this path, by P, we shall have for the corresponding quantity of work, Pp, and for the entire quantity of work denoted by Q, we shall have the sum of these elementary quantities of work; or, since p is the same for each,

$$Q = p\, \Sigma(P) \quad \ldots \quad (134.)$$

The quotient obtained by dividing the entire quantity of work by the entire path, is called the *mean pressure*, or the *mean resistance*, and is evidently the force which, acting uniformly through the same path, would accomplish the same work.

Work, when the power acts obliquely to the path.

144. Let PD represent the force, and AB the path which the body D is constrained to follow. Denote the angle PDs by α, and suppose P to be resolved into two components, one perpendicular, and the other parallel to AB. We shall have, for the former, $P\sin\alpha$, and, for the latter, $P\cos\alpha$.

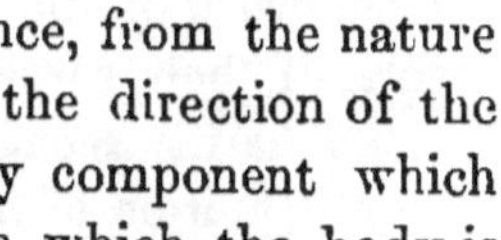

Fig. 126.

The former can produce no work, since, from the nature of the case, the point cannot move in the direction of the normal; hence, the latter is the only component which works. Let sD be the space through which the body is moved in any time whatever. If we denote the pressure exerted in the direction of PD, by P, and the quantity of work, by Q, we shall have,

$$Q = P\cos\alpha \times \overline{sD}.$$

Let fall the perpendicular ss' from s, on the direction of the force P. From the right-angled triangle Dss', we shall have,

$$sD \times \cos a = s'D.$$

Substituting this in the preceding equation, we get,

$$Q = P \times s'D.$$

That is, the quantity of work of a force acting obliquely to the path along which the point of application is constrained to move, is equal to the intensity of the force multiplied by the projection of the path upon the direction of the force. We have supposed the intensity of the force P, to be expressed in pounds, or units of mass.

If we take the distance sD, infinitely small, $s'D$ will be the virtual velocity of D, and the expression for the quantity of work of P will be its virtual moment (Art. 38). Hence we say that the elementary quantity of work of a force is equal to its virtual moment, and, from the principle of virtual moments, we conclude that the algebraic sum of the elementary quantities of work of any number of forces applied at the same point, is equal to the elementary quantity of work of their resultant. What is true for the elementary quantities of work at any instant, must be equally true at any other instant. Hence, the algebraic sum of all the elementary quantities of work of the components in any time whatever, is equal to the algebraic sum of the elementary quantities of work of their resultant for the same time; that is, the work of the components for any time, is equal to the work of their resultant for the same time. This principle would hardly seem to require demonstration, for, from the very definition of a resultant, it would seem to be true of necessity. If the forces are in equilibrium, the entire quantity of work will be equal to 0.

This principle finds an important application, in computing the quantity of work required to raise the material for a wall or building; for raising the material from a shaft; for raising water from one reservoir to another; and a great variety of similar operations. In this connection, the principle may be enunciated as follows: *The algebraic sum of the quantities of work required to raise the parts of a system through any vertical spaces, is equal to the quantity of work required to move the whole system over a vertical space equal to that described by the centre of gravity of the system.*

It also follows, from the same principle, that, *if all the pieces of a machine which moves without friction be in equilibrium in all positions, under the action of weights suspended from different parts of the machine, the centre of gravity of the system will neither ascend nor descend whilst the machine is in motion.*

Work, when a body is constrained to move upon a curve.

145. Let AB represent the curve, and suppose that the force is so taken that its line of direction shall always pass through a point P. Divide the curve into elements so small that each may be taken as a straight line, and, with P as a centre, and the distances from P to the points of division as radii, describe arcs of circles. Then, denoting the force supposed constant, by P, we shall have (from Art. 144) the elementary quantity of work performed whilst the point is moving over aa', equal to $P \times ac$, or $P \times bb'$. In like manner, the quantity of work performed whilst the point is describing $a'a''$ will be equal to $P \times b'b''$, and so on. Hence, by summation, we shall find the entire quantity of work performed in moving the body from B to A will be equal to $P \times BB'$. If now we suppose the curve AB to lie in a vertical plane, and the force to be the force of gravity, the point P may be regarded as infinitely distant, the lines Pa, Pa' &c., will become vertical, and the lines $a'b'$, $a''b''$, will be horizontal. We may, therefore, enunciate the following principle: The quantity of work of the weight of a body

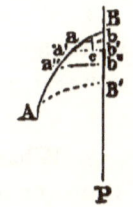

Fig. 127.

in descending a curve, is equal to the quantity of work of the same weight in descending vertically through the same height. This principle is immediately connected with the discussion in Art. 74.

If a body in a stable position, as a pyramid resting on its base, be overturned by any extraneous force, the quantity of work will be equal to the weight of the body, multiplied by the vertical height to which the centre of gravity must be raised before reaching its highest point. This product might be taken as the measure of the stability of a body.

EXAMPLES.

1. What amount of work is required to raise 500 lbs. to the height of 5 yards? *Ans.* 7500 units, or 7500 lbs. ft.

2. To what height can 2240 lbs. be raised by the expenditure of 5600 units of work? *Ans.* 2.5 ft.

3. What weight can be raised to the height of 25 feet by 224000 units of work? *Ans.* 8960 lbs.

4. What is the effective horse power of an engine which raises 80 cubic feet of water per minute from the depth of 360 feet, a cubic foot of water weighing 62 lbs.
Ans. 54.11 horse power.

5. What must be the effective horse power to raise the same quantity of water per minute, from a depth of 40 feet?
Ans. 6 horse power.

6. How many tons of ore can be raised per hour from a mine 1800 feet deep, by an engine of 28 effective horse power, reckoning 2240 lbs. to the ton? *Ans.* $13\frac{3}{4}$ tons.

7. From what depth will an engine of 16 effective horse power raise 5 cwts. of coal per minute.
Ans. 943 feet, nearly.

8. In what time will an engine of 40 effective horse power raise 44000 cubic feet of water from a mine 360 feet deep, allowing $62\frac{1}{2}$ pounds to the cubic foot?
Ans. 12 h. 30 min.

CURVILINEAR AND ROTARY MOTION. 221

9. Required the quantity of work necessary to raise the material for a rectangular granite wall 25 feet long, $2\frac{1}{2}$ feet thick, and 20 feet high, the weight of granite being 162 lbs. per cubic foot?

SOLUTION.

The weight of the wall is equal to

162 lbs. × 25 × 2.5 × 20 = 202500 lbs.

The height of the centre of gravity being 10 feet, the quantity of work is equal to

202500 × 10 = 2025000 lbs. ft. *Ans.*

10. How long would it take an engine of 4 effective horse power to raise the material for the wall in the last example? *Ans.* $15\frac{1}{4}$ minutes, nearly.

11. What quantity of work must be expended in drawing a chain from a shaft, the length of the chain being 450 feet, and its weight 40 lbs. to the foot? *Ans.* 4050000 lbs. ft.

12. A cylindrical well is 150 feet deep, and 10 feet in diameter. Supposing the well to be filled with water to the depth of 50 feet, how much work must be expended in raising it to the top, water being taken at 62.5 lbs. per cubic foot?

SOLUTION.

The weight of the water is equal to

$\pi \times 5^2 \times 50 \times 62.5$ lbs. = 245437.5 lbs.

The distance of the centre of gravity from the top is 125 feet. Hence, the required quantity of work is equal to

245437.5 lbs. × 125 ft. = 30679687.5 lbs. ft. *Ans.*

13. What quantity of work will be required to overturn a right cone, with a circular base, whose altitude is 12

feet, and the radius of whose base is 4 feet, the weight of the material being estimated at 100 lbs. per cubic foot?

SOLUTION.

The weight of the cone is equal to

$$\pi \times 4^2 \times 4 \times 100 \text{ lbs.} = 20106.24 \text{ lbs.}$$

If the cone turns about a tangent to its base, since the centre of gravity is 3 feet from the base, it will be,

$$\sqrt{3^2 + 4^2} = 5 \text{ feet from the tangent.}$$

The centre of gravity, at its highest point, will, therefore, be 5 feet from the horizontal plane. It must then be raised 2 feet. Hence, the required quantity of work is equal to

$$20106.24 \text{ lbs.} \times 2 \text{ ft.} = 40212.48 \text{ lbs. ft.} \quad Ans.$$

14. To show that the work required for overturning similar solids, similarly placed, varies as the fourth powers of their homologous lines.

SOLUTION.

Denote the altitudes of the centres of gravity, by y and ry, the distances from the directions of the weights to the lines about which they turn, by x and rx, and their weights, by w and r^3w.

The quantity of work required to overturn the first, will be,

$$Q = w(\sqrt{x^2 + y^2} - y).$$

The quantity of work required to overturn the second, will be,

$$Q' = r^3w(\sqrt{r^2x^2 + r^2y^2} - ry) = r^4w(\sqrt{x^2 + y^2} - y).$$

Hence,

$$Q : Q' :: 1 : r^4 \quad . \quad . \quad Q.E.D.$$

Rotation.

146. When a body restrained by a fixed axis, about which it is free to turn, is acted upon by a force, it will, in general, take up a motion of *rotation*, or *revolution*. In this kind of motion, each point of the body describes a circle, whose centre is in the axis, and whose plane is perpendicular to the axis. The time of a complete revolution being the same for each particle, it follows, that the velocities of the different particles will be proportional to their distances from the axis. The velocity of any particle will be equal to its distance from the axis multiplied by the angular velocity (Art. 122).

Quantity of work of a Force producing Rotation.

147. If a force is applied obliquely to the axis of rotation, we may conceive it to be resolved into two components, one parallel, and the other perpendicular to the axis of rotation. The effect of the former will be counteracted by the resistance offered by the fixed axis; the effect of the latter in producing rotation will be exactly the same as that of the applied force. We need, therefore, only consider those components whose directions are perpendicular to the axis of rotation.

Let P represent any force whose line of direction is perpendicular to the axis, but does not intersect it. Let O be the point in which a plane through P, perpendicular to the axis, intersects it. Let A and C be any two points whatever, on the line of direction of P. Suppose the

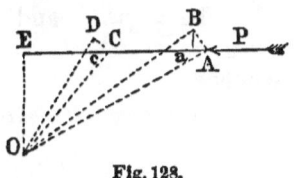

Fig. 128.

force P to turn the system through an infinitely small angle, and let B and D be the new positions of A and C. Draw OE, Ba, and Dc respectively perpendicular to PE; draw also, AO, BO, CO, and DO. Denote the distances OA, by r, OC, by r', OE, by p, and the path described by a point at a unit's distance from O, by θ'. Since the angles AOB, and COD are equal, from the nature of the motion of rotation, we shall have, $AB = r\theta'$, and $CD = r'\theta'$; and since the angular motion is infinitely small, these lines may be regarded as straight lines, perpendicular respectively to OA and OC. From the right-angled triangles ABa and CDc, we have,

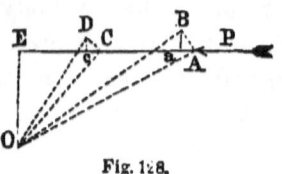

Fig. 128.

$$Aa = r\theta' \cos BAa, \quad \text{and} \quad Cc = r'\theta' \cos DCc.$$

In the right-angled triangles ABa, and OAE, we have AB perpendicular to OA, and Aa perpendicular to OE; hence, the angles BAa, and AOE, are equal, as are also their cosines; hence, we have,

$$\cos BAa = \cos AOE = \frac{p}{r}.$$

In like manner, it may be shown, that

$$\cos DCc = \cos COE = \frac{p}{r'}.$$

Substituting in the equations just deduced, we have,

$$Aa = p\theta, \quad \text{and} \quad Cc = p\theta ; \quad \therefore \ Aa = Cc ;$$

whence,

$$P \cdot Aa = P \cdot Cc = Pp\,\theta'.$$

The first member of the equation is this quantity of work of P, when its point of application is at A ; the second is the quantity of work of P, when its point of application is at C. Hence, we conclude, that *the elementary quantity of work of a force applied to produce rotation, is always the*

CURVILINEAR AND ROTARY MOTION. 225

same, wherever its point of application may be taken, provided its line of direction remains unchanged.

We conclude, also, that the elementary quantity of work is equal to the intensity of the force multiplied by its lever arm into the elementary space described by a point at a unit's distance from the axis.

If we suppose the force to act for a unit of time, the intensity and lever arm remaining the same, and denote the *angular velocity*, by θ, we shall have,

$$Q' = Pp\theta.$$

For any number of forces similarly applied, we shall have,

$$Q = \Sigma(Pp)\theta \quad \ldots \quad (135.)$$

If the forces are in equilibrium, we shall have (Art. 49), $\Sigma(Pp) = 0$; consequently, $Q = 0$.

Hence, if any number of forces tending to produce rotation about a fixed axis, are in equilibrium, the entire quantity of work of the system of forces will be equal to 0.

Accumulation of Work.

148. When a body is put in motion by the action of a force, its inertia has to be overcome, and, in order to bring the body back again to a state of rest, a quantity of work has to be given out just equal to that required to put it in motion. This results from the nature of inertia. A body in motion may, therefore, be regarded as the representation of a quantity of work which can be reproduced upon any resistance opposed to its motion. Whilst the body is in motion, the work is said to be *accumulated*. In any given instance, the *accumulated work* depends, *first*, upon the mass in motion; and, *secondly*, upon the velocity with which it moves.

Take the case of a body projected vertically upwards in vacuum. The projecting force expends upon the body a quantity of work sufficient to raise it through a height equal

10*

to that due to the velocity of projection. Denoting the weight of the body, by w, the height to which it rises, by h, and the accumulated work, by Q, we shall have,

$$Q = wh.$$

But, $h = \tfrac{1}{2}\dfrac{v^2}{g}$, (Art. 116), hence,

$$Q = \tfrac{1}{2}\dfrac{w}{g}v^2.$$

Denoting the mass of the body by m, we shall have, $m = \dfrac{w}{g}$ (Art. 11), and, by substitution, we have, finally,

$$Q = \tfrac{1}{2}mv^2 \quad \ldots \ldots \quad (136.)$$

If the body descends by its own weight, it will have impressed upon it by the force of gravity, during the descent, exactly the same quantity of work as it gave out in ascending.

The amount of work accumulated in a body is evidently the same, whatever may have been the circumstances under which the velocity has been acquired; and also, the amount of work which it is capable of giving out in overcoming any resistance is the same, whatever may be the nature of that resistance. Hence, *the measure of the accumulated work of a moving mass is one-half of the mass into the square of the velocity.*

The expression mv^2, is called the *living force* of the body. Hence, the *living force of a body is equal to its mass, multiplied by the square of its velocity.* The living force of a body is the measure of twice the quantity of work expended in producing the velocity, or, it is the measure of twice the quantity of work which the body is capable of giving out.

When the forces exerted tend to increase the velocity,

their work is regarded as positive; when they tend to diminish it, their work is regarded as negative. It is the aggregate of all the work expended, both positive and negative, that is measured by the quantity, $\frac{1}{2}mv^2$.

If, at any instant, a body whose mass is m, has a velocity v, and, at any subsequent instant, its velocity has become v', we shall have, for the accumulated work at these two instants,

$$Q = \tfrac{1}{2}mv^2, \quad Q' = \tfrac{1}{2}mv'^2;$$

and, for the aggregate quantity of work expended in the interval,

$$Q'' = \tfrac{1}{2}m(v'^2 - v^2) \quad \ldots \quad (137.)$$

When the motive forces, during the interval, perform a greater quantity of work than the resistances, the value of v' will be greater than that of v, and there will be an accumulation of work in the interval. When the work of the resistances exceeds that of the motive forces, the value of v will exceed that of v', Q'' will be negative, and there will be a loss of living force, which is absorbed by the resistances.

Living Force of Revolving Bodies.

149. Denote the angular velocity of a body which is restrained by an axis, by θ; denote the masses of its elementary particles by m, m', &c., and their distances from the axis of rotation, by r, r', &c. Their velocities will be $r\theta$, $r'\theta$, &c., and their living forces will be $mr^2\theta^2$, $m'r'^2\theta^2$, &c. Denoting the entire living force of the body, by L, we shall have, by summation, and recollecting that θ^2 is the same for all the terms,

$$L = \Sigma(mr^2)\theta^2 \quad \ldots \quad (138.)$$

But $\Sigma(mr^2)$ is the expression for the moment of inertia of the body, taken with respect to the axis of rotation. Denoting the entire mass by M, its radius of gyration, with respect to the axis of rotation, by k, we shall have,

$$L = Mk^2\theta^2.$$

If, at any subsequent instant, the angular velocity has become θ', we shall, at that instant, have,

$$L' = Mk^2\theta'^2\,;$$

and, for the loss or gain of living force in the interval, we shall have,

$$L'' = Mk^2(\theta'^2 - \theta^2). \quad . \quad . \quad (139.)$$

If we make $\theta'^2 - \theta^2 = 1$, we shall have,

$$L''' = Mk^2 = \Sigma(mr^2) \quad . \quad . \quad (140.)$$

which shows that the moment of inertia of a body, with respect to an axis, is equal to the living force lost or gained whilst the body is experiencing a change in the square of its angular velocity equal to 1.

The principle of living forces is extensively applied in discussing the circumstances of motion of machines. When the motive power performs a quantity of work greater than that necessary to overcome the resistances, the velocities of the parts become accelerated, a quantity of work is stored up, to be again given out when the resistances offered require a greater quantity of work to overcome them than is furnished by the motor.

In many machines, pieces are expressly introduced to equalize the motion, and this is particularly the case when either the motive power or the resistance to be overcome, is, in its nature, variable. Such pieces are called *fly-wheels*.

Fly-Wheels.

150. A fly-wheel is a heavy wheel, usually of iron, mounted upon an axis, near the point of application of the force which it is destined to regulate. It is generally composed of a heavy rim, connected with the axis by means of radial arms. Sometimes it consists of radiating bars, carrying heavy spheres of metal at their outer extremity. In either case, we see, from Equation 139, that, for a given quantity of work absorbed, the value of $\theta'^2 - \theta^2$ will be less as M and k are greater; that is, the change of angular velocity will be less, as the mass of the fly-wheel and its radius of gyration increase.

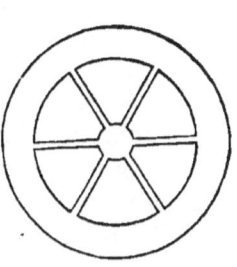

Fig. 129.

It is for this reason that the peculiar form of fly-wheel indicated above, is adopted, it being the form that most nearly realizes the conditions pointed out. The principal objection to large fly-wheels in machinery, is the great amount of hurtful resistance which they create, such as friction on the axle, &c. Thus, a fly-wheel of 42000 lbs. would create a force of friction of 4200 lbs., the coefficient of friction being but $\frac{1}{10}$; and, if the diameter of the axle were 8 inches, and the number of revolutions 30 per minute, this resistance alone would be equal to 8 horse powers.

EXAMPLES.

1. The weight of the ram of a pile-driver is 400 lbs., and it strikes the head of a pile with a velocity of 20 feet. What is the amount of work stored up in it?

SOLUTION.

The height due to the velocity, 20 feet, is equal to

$$\frac{400}{64\frac{1}{3}} = 6.22 \text{ ft., nearly.}$$

Hence, the stored up work is equal to

$$400 \text{ lbs.} \times 6.22 \text{ ft.} = 2488 \text{ lbs. ft.};$$

230 MECHANICS.

or, the stored up work, equal to half the living force, is equal to

$$\frac{400}{32\frac{1}{6}} \times \frac{(20)^2}{2} = 2488 \text{ units. } Ans.$$

2. A train, weighing 60 tons, has a velocity of 40 miles per hour when the steam is shut off. How far will it travel, if no brake be applied, before the velocity is reduced to 10 miles per hour, the resistance to motion being estimated at 10 lbs. per ton. *Ans.* 11236 ft.

Composition of Rotations.

151. Let a body $ACBD$, that is free to move, be acted upon by a force which, of itself, would cause the body to revolve for the infinitely small time dt, about the line AB, with an *angular velocity* v; and at the same instant, let the body be acted upon by a second force, which would of itself cause the body to revolve about CD, for the time dt, with an *angular velocity* v'.

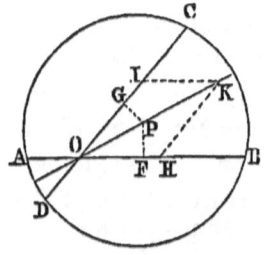

Fig. 130.

Suppose the axes to intersect each other at O, and let P be any point in the plane of the axes. Draw PF and PG respectively perpendicular to OB and OC, denoting the former, by x, and the latter, by y. Then will the velocity of P due to the first force, be equal to vx, and its velocity due to the second force will be equal to $v'y$. Suppose the rotation to take place in such a manner, that the tendency of the rotation about one of the axes, shall be to depress the point below the plane, whilst that about the other is to elevate it above the plane; then will the effective velocity of P be equal to $vx - v'y$. If this effective velocity is 0, *the point P will remain at rest*. Placing the expression just deduced equal to 0, and transposing, we have,

$$vx = v'y.$$

CURVILINEAR AND ROTARY MOTION. 231

To determine the position of P, lay off OH equal to v, OI equal to v', and regard these lines as the representatives of two forces; we have, from the equation, the moment of v, with respect to the point P, equal to the moment of v', with respect to the same point. Hence, the point P must be somewhere upon the diagonal OK, of the parallelogram described on v, and v'. But P may be anywhere on this line; hence, every point of the diagonal OK, remains at rest during the time dt, and is, consequently, *the resultant axis of rotation*. We have, therefore, the following principles:

If a body be acted upon simultaneously by two forces, each tending to impart a motion of rotation about a separate axis, the resultant motion will be one of rotation about a third axis lying in the plane of the other two, and passing through their common point of intersection.

The direction of the resultant axis coincides with the diagonal of a parallelogram, whose adjacent sides are the component axes, and whose lengths are proportional to the impressed angular velocities.

Let OH and OI represent, as before, the angular velocities v and v', and OK the diagonal of the parallelogram constructed on these lines as sides. Take any point I, on the second axis, and let fall a perpendicular on OH and OK; denote the former by r, and the latter, by r''; denote, also, the resultant angular velocity, by v''.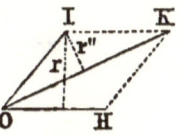

Fig. 131.

Since the actual space passed over by I, during the time t, depends only upon the first force, it will be the same whether we regard the revolution as taking place about the axis OH, or about the axis OK. If we suppose the rotation to take place about OH, the space passed over in the time dt, will be equal to $rvdt$; if we suppose the rotation to take place about OK, the space passed over in the same time will be equal to $r''v''dt$. Placing these expressions equal to each other, we have, after reduction,

$$v'' = \frac{r}{r''} v.$$

But regarding I as a centre of moments, we shall have, from the principle of moments,

$$OK \times r'' = vr; \quad \text{or,} \quad OK = \frac{r}{r''} v.$$

By comparing the last two equations, we have,

$$v'' = OK.$$

That is, *the resultant angular velocity will be equal to the diagonal of the parallelogram described on the component angular velocities as sides.*

By a course of reasoning entirely similar to that employed in demonstrating the parallelopipedon of forces, we might show, that,

If a body be acted upon by three simultaneous forces, each tending to produce rotation about separate axes intersecting each other, the resultant motion will be one of rotation about the diagonal of the parallelopipedon whose adjacent edges are the component angular velocities, and the resultant angular velocity will be represented by the length of this diagonal.

The principles just deduced are called, respectively, *the parallelogram* and *the parallelopipedon of rotations.*

Application to the Gyroscope.

152. The gyroscope is an instrument used to illustrate the laws of rotary motion. It consists essentially of a heavy wheel A, mounted upon an axle BC. This axle is attached, by means of pivots, to the inner edge of a circular hoop DE, within which the wheel A can turn freely. On

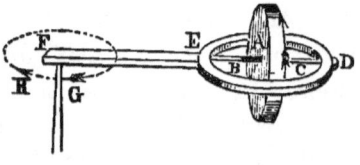

Fig. 132.

one side of the hoop, and in the prolongation of the axle BC, is a bar EF, having a conical hole drilled on its lower face to receive the pointed summit of a vertical standard G. If a string be wrapped several times around the axle BC, and then rapidly unwound, so as to impart a rapid motion of rotation to the wheel A, in the direction indicated by the arrow-head, it is observed that the machine, instead of sinking downwards under the action of gravity, takes up a retrograde orbital motion about the pivot G, as indicated by the arrow-head H. For a time, the orbital motion increases, and, under certain circumstances, the bar EF is observed to rise upwards in a retrograde spiral direction; and, if the cavity for receiving the pivot is pretty shallow, the bar may even be thrown off the vertical standard. Instead of a bar EF, the instrument may simply have an ear at E, and be suspended from a point above by means of a string attached to the ear. The phenomena observed are the same as before.

Before explaining these phenomena, it will be necessary to point out the conventional rules for attributing proper signs to the different rotations.

Let OX, OY, and OZ, be three rectangular axes. It has been agreed to call all distances, estimated from O, towards either X, Y, or Z, *positive*; consequently, all distances estimated in a contrary direction must be regarded as negative. If a body revolve about either axis, or about any line through the origin, in such a manner as to appear to an eye beyond it, in the axis and looking towards the origin, to move in the same direction as the hands of a watch, that rotation is considered *positive*. If rotation takes place in an opposite direction, it is *negative*. The arrow-head A, indicates the direction of positive rotation about the axis of X. To an eye situated beyond the body, as at X, and looking towards the origin, the motion appears to be in the same direction as the motion of the hands of a watch. The arrowhead B,

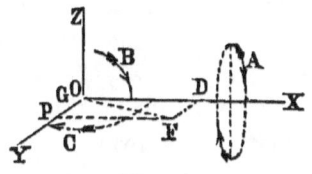

Fig. 133.

indicates the direction of positive rotation about the axis of Y, and the arrow-head C, the direction of positive rotation about the axis of Z.

Suppose the axis of the wheel of the gyroscope to coincide with the axis of X, taken horizontal; let the standard be taken to coincide with the axis of Z, the axis of Y being perpendicular to them both. Let a positive rotation be communicated to the wheel by means of a string. For a very short time dt, the angular velocity may be regarded as constant. In the same time dt, the force of gravity acts to impart a motion of positive rotation to the whole instrument about the axis of Y, which may, for an instant, be regarded as constant. Denote the former angular velocity by v, and the latter by v'. Lay off in a positive direction on the axis of X, the distance OD equal to v, and, on the positive direction of the axis of Y, the distance OP equal to v', and complete the parallelogram OF. Then (Art. 151) will OF represent the direction of the resultant axis of revolution, and the distance OF will represent the resultant angular velocity, which denote by v''. In moving from OD to OF, the axis takes up a positive, or *retrograde orbital motion* about the axis of Z. To construct the position of the resultant axis for the second instant dt, we must compound three angular velocities. Lay off on a perpendicular to OF and OZ, the angular velocity OG due to the action of gravity during the time dt, and on OZ the angular velocity in the orbit; construct a parallelopipedon on these lines, and draw its diagonal through O. This diagonal will coincide in direction with the resultant axis for the second instant, and its length will represent the resultant angular velocity (Art. 151). For the next instant, we may proceed as before, and so on continually. Since, in each case, the diagonal is greater than either edge of the parallelopipedon, it follows that the angular velocity will continually increase, and, were there no hurtful resistances, this increase would go on indefinitely. The effect of gravity is continually exerted to depress the centre of gravity of the instrument, whilst the effect of the orbital rotation is to elevate it. When the latter effect prevails, the axis of the gyroscope will continually rise; when the former prevails, the gyroscope will continually descend. Whether the one or the other of these conditions will be fulfilled, depends upon the angular velocity of the wheel of the gyroscope, and upon the position of the centre of gravity of the instrument. Were the instrument counterpoised so that the centre of gravity would lie exactly over the pivot, there would be no orbital motion, neither would the instrument rise or fall. Were the centre of gravity thrown on the opposite side of the pivot from the wheel, the rotation due to gravity would be negative, that is, the orbital motion would be *direct*, instead of *retrograde*.

CHAPTER VII.

MECHANICS OF LIQUIDS.

Classification of Fluids.

153. A FLUID is a body whose particles move freely amongst each other, each particle yielding to the slightest force. Fluids are of two classes: *liquids*, of which water is a type, and *gases*, or *vapors*, of which air and steam are types. The distinctive property of the first class is, that they are sensibly *incompressible*; thus, water, on being pressed by a force of 15 lbs. on each square inch of surface, only suffers a diminution of about $\frac{1}{200000}$ of its bulk. The second class comprises those which are readily compressible; thus, air and steam are easily compressed into smaller volumes, and when the pressure is removed, they expand, so as to occupy larger volumes.

Most liquids are imperfect; that is, there is more or less adherence between their particles, giving rise to viscosity. In what follows, they will be regarded as destitute of *viscosity*, and *homogeneous*. For certain purposes, fluids may also be regarded as destitute of weight, without impairing the validity of the conclusions.

Principle of Equal Pressures.

154. From the nature and constitution of a fluid, it follows, that each of its particles is perfectly movable in all directions. From this fact, we deduce the following fundamental law, viz.: *If a fluid is in equilibrium under the action of any forces whatever, each particle of the mass is equally pressed in all directions;* for, if any particle were more strongly pressed in one direction than in the others, it would yield in that direction, and motion would ensue, which is contrary to the hypothesis.

This is called the *principle of equal pressures.*

It follows, from the principle of equal pressures, that if any point of a fluid in equilibrium, be pressed by any force, that pressure will be transmitted without change of intensity to every other point of the fluid mass.

This may be illustrated experimentally, as follows:

Let AB represent a vessel filled with a fluid in equilibrium. Let C and D represent two openings, furnished with tightly-fitting pistons. Suppose that forces are applied to the pistons just sufficient to maintain the fluid mass in equilibrium. If, now, any additional force be applied to the piston P, the piston Q will be forced outwards; and in order to prevent this, and restore the equilibrium, it will be found necessary to apply a force to the piston Q, which shall have the same ratio to the force applied at P that the area of the piston Q has to the area of the piston P. This principle will be found to hold true, whatever may be the sizes of the two pistons, or in whatever portions of the surface they may be inserted. If the area of P be taken as a unit, then will the pressure upon Q be equal to the pressure on P, multiplied by the area of Q.

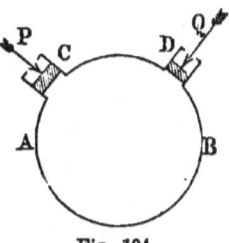

Fig. 134.

The pressure transmitted through a fluid in equilibrium, to the surface of the containing vessel, is normal to that surface; for if it were not, we might resolve it into two components, one normal to the surface, and the other tangential; the effect of the former would be destroyed by the resistance of the vessel, whilst the latter would impart motion to the fluid, which is contrary to the supposition of equilibrium.

In like manner, it may be shown, that the resultant of all the pressures, acting at any point of the free surface of a fluid, is normal to the surface at that point. When the only force acting is the force of gravity, the surface is level. For small areas, a level surface coincides sensibly with a horizontal plane. For larger areas, as lakes and oceans, a level surface coincides with the general surface of the earth. Were the earth at rest, the level surface of lakes and oceans would be spherical; but, on account of the centrifugal force arising from the rotation of the earth, it is sensibly an ellipsoidal surface, whose axis of revolution is the axis of the earth.

Pressure due to Weight.

155. If an incompressible fluid be in a state of equilibrium, the pressure at any point of the mass arising from the weight of the fluid, is proportional to the depth of the point below the free surface.

Take an infinitely small surface, supposed horizontal, and conceive it to be the base of a vertical prism whose altitude is equal to its distance below the free surface. Conceive this filament to be divided by horizontal planes into infinitely small, or elementary prisms. It is evident, from *the principle of equal pressures*, that the pressure upon the lower face of any one of these elementary prisms is greater than that upon its upper face, by the weight of the element, whilst the lateral pressures are such as to counteract each other's effects. The pressure upon the lower face of the first prism, counting from the top, is, then, just equal to its weight; that upon the lower face of the second is equal to the weight of the first, *plus* the weight of the second, and so on to the bottom. Hence, the pressure upon the assumed surface is equal to the weight of the entire column of fluid above it. Had the assumed elementary surface been oblique to the horizon, or perpendicular to it, and at the same depth as before, the pressure upon it would have been the same, from *the principle of equal pressures*. We have, therefore, the following law:

The pressure upon any elementary portion of the surface of a vessel containing a heavy fluid is equal to the weight of a prism of the fluid whose base is equal to that surface, and whose altitude is equal to its depth below the free surface.

Denoting the area of the elementary surface, by s, its depth below the free surface, by z, the weight of a unit of the volume of the fluid, by w, and the pressure, by p, we shall have,

$$p = wzs \quad \ldots \quad \ldots \quad (141.)$$

We have seen that the pressure upon any element of a surface is normal to the surface. Denote the angle which this normal makes with the vertical, estimated from above, downwards, by φ, and resolve the pressure into two components, one vertical and the other horizontal, denoting the vertical component by p', we shall have,

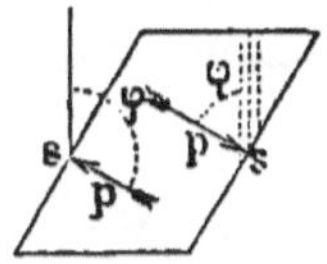

Fig. 135.

$$p' = wzs\cos\varphi \quad \ldots \quad \ldots \quad (142.)$$

But $s\cos\varphi$ is equal to the horizontal projection of the elementary surface s, or, in other words, it is equal to a horizontal section of a vertical prism, of which that surface is the base. Hence, *the vertical component of the pressure on any element of the surface is equal to the weight of a column of the fluid, whose base is equal to the horizontal projection of the element, and whose altitude is equal to the distance of the element from the upper surface of the fluid.*

The distance z has been estimated as *positive* from the surface of the fluid downwards. If $\varphi < 90°$, we have $\cos\varphi$ positive; hence, p' will be *positive*, which shows that the vertical pressure is exerted downwards. If $\varphi > 90°$, we have $\cos\varphi$ *negative;* hence, p' is negative, which shows that the vertical pressure is exerted upwards (see Fig. 135).

Suppose the interior surface of a vessel containing a heavy fluid to be divided into elementary portions, whose areas are denoted by s, s', s'', &c.; denote the distances of these elements below the upper surface, by z, z', z'', &c. From the principle just demonstrated, the pressures upon these surfaces will be denoted by wsz, $ws'z'$, $ws''z''$, &c., and the entire pressure upon the interior of the vessel will be equal to,

$$w(sz + s'z' + s''z'' + \&c.); \quad \text{or,} \quad w \times \Sigma(sz).$$

Let Z denote the depth of a column of the fluid, whose base is equal to the entire surface pressed, and whose weight is equal to the entire pressure, then will this pressure be equal to $w(s + s' + s'' + \&c.)Z$; or, $wZ \cdot \Sigma s$. Equating these values, we have,

$$w \cdot \Sigma(sz) = wZ \cdot \Sigma(s), \quad \therefore \quad Z = \frac{\Sigma(sz)}{\Sigma(s)} \quad . \quad (143.)$$

The second member of (143), (Art. 51), expresses the distance of the centre of gravity of the surface pressed, below the free surface of the fluid. Hence,

The entire pressure of a heavy fluid upon the interior of the containing vessel, is equal to the weight of a volume of the fluid, whose base is equal to the area of the surface pressed, and whose altitude is equal to the distance of the centre of gravity of the surface from the free surface of the fluid.

EXAMPLES.

1. A hollow sphere is filled with a liquid. How does the entire pressure, on the interior surface, compare with the weight of the liquid?

SOLUTION.

Denote the radius of the interior surface of the sphere, by r, and the weight of a unit of volume of the liquid, by w. The entire surface pressed is measured by $4\pi r^2$; and, since the centre of gravity of the surface pressed is at a distance r below the surface of the liquid, the entire pressure on the interior surface will be measured by the expression,

$$w \times 4\pi r^2 \times r = 4\pi w r^3.$$

But the weight of the liquid is equal to

$$\tfrac{4}{3}\pi w r^3.$$

Hence, *the entire pressure is equal to three times the weight of the liquid.*

2. A hollow cylinder, with a circular base, is filled with a liquid. How does the pressure on the interior surface compare with the weight of the liquid?

SOLUTION.

Denote the radius of the base of the cylinder, by r, and the altitude, by h. The centre of gravity of the lateral surface is at a distance below the upper surface of the fluid equal to $\tfrac{1}{2}h$. If we denote the weight of the unit of volume of the liquid, by w, we shall have, for the entire pressure on the interior surface,

$$wh\pi r^2 + 2w\pi r \cdot \tfrac{1}{2}h^2 = w\pi r h(r + h).$$

But the weight of the liquid is equal to

$$w\pi r^2 h.$$

Hence, *the total pressure is equal to $\dfrac{r+h}{r}$ times the weight of the liquid.*

If we suppose $h = r$, the pressure will be twice the weight.

If we suppose $r = 2h$, we shall have the pressure equal to $\tfrac{3}{2}$ of the weight.

If we suppose $h = 2r$, the pressure will be equal to three times the weight, and so on.

In all cases, the total pressure will exceed the weight of the liquid.

3. A right cone, with a circular base, stands on its base, and is filled with a liquid. How does the pressure on the internal surface compare with the weight of the liquid?

SOLUTION.

Denote the radius of the base, by r, and the altitude, by h, then will the slant height be equal to

$$\sqrt{h^2 + r^2}.$$

The centre of gravity of the lateral surface, below the upper surface of the liquid is equal to $\frac{2}{3}h$. If we denote the weight of a unit of volume of the liquid, by w, we shall have, for the total pressure on the interior surface,

$$w\pi r^2 h + \tfrac{2}{3} w\pi r h \sqrt{h^2 + r^2} = w\pi r h (r + \tfrac{2}{3}\sqrt{h^2 + r^2}).$$

But the weight of the liquid is equal to

$$\tfrac{1}{3} w\pi r^2 h = w\pi r h \times \tfrac{1}{3} r.$$

Hence, *the total pressure is equal to* $\dfrac{3r + 2\sqrt{h^2 + r^2}}{r}$ *times the weight.*

4. Required the relation between the pressure and the weight in the preceding case, when the cone stands on its vertex.

SOLUTION.

The total pressure is equal to

$$\tfrac{1}{3} w\pi r h \sqrt{h^2 + r^2};$$

and, consequently, *the pressure is equal to* $\dfrac{\sqrt{h^2 + r^2}}{r}$ *times the weight of the liquid*

MECHANICS OF LIQUIDS.   243

5. What is the pressure on the lateral faces of a cubical vessel filled with water, the edges of the cube being 4 feet, and the weight of the water 62½ lbs. per cubic foot?

Ans. 8000 lbs.

6. A cylindrical vessel is filled with water. The height of the vessel is 4 feet, and the radius of the base 6 feet. What is the pressure on the lateral surface?

Ans. 18850 lbs., nearly.

Centre of Pressure on a Plane Surface.

156. Let $ABCD$ represent a plane, pressed by a fluid on its upper surface, AB its intersection with the free surface of the fluid, G its centre of gravity, O the centre of pressure, and s the area of any element of the surface at S. Denote the inclination of the plane to the level surface, by α, the perpendicular distances from O to AB, by x, from G to AB, by p, and from S to AB, by r. Denote, also, the entire area AC, by A, and the weight of a unit of volume of the fluid, by w. The perpendicular distance from G to the free surface of the fluid, will be equal to $p \sin\alpha$, and that of any element of the surface, will be $r \sin\alpha$.

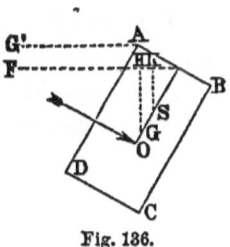

Fig. 136.

From the preceding article, it follows that the entire pressure exerted is equal to $wAp \sin\alpha$, and its moment, with respect to AB as an axis of moments, is equal to

$$wAp \sin\alpha \times x.$$

The elementary pressure on s is, in like manner, equal to $wsr \sin\alpha$, and its moment, with respect to AB, is $wsr^2 \sin\alpha$, and the sum of all the elementary moments is equal to

$$w \sin\alpha \, \Sigma(sr^2).$$

But the resultant moment is equal to the algebraic sum of the elementary moments. Hence,

$$wAp \sin\alpha \times x = w \sin\alpha \, \Sigma(sr^2);$$

and, by reduction,

$$x = \frac{\Sigma(sr^2)}{Ap} \quad \ldots \ldots \quad (144.)$$

The numerator is the moment of inertia of the plane $ABCD$, with respect to AB, and the denominator is the moment of the area with respect to the same line. Hence, *the distance from the centre of pressure to the intersection of the plane with the free surface, is equal to the moment of inertia of the plane, divided by the moment of the plane.*

If we take the straight line AD, perpendicular to AB, as an axis of moments, denoting the distance of O from it, by y, and of s from it, by l, we shall, in a similar manner, have,

$$wAp \sin\alpha\, y = w\sin\alpha\, \Sigma(srl);$$

and, by reduction,

$$y = \frac{\Sigma(srl)}{Ap}. \quad \ldots \ldots \quad (145.)$$

The values of x and y make known the position of the centre of pressure.

EXAMPLES.

1. What is the position of the centre of pressure on a rectangular flood-gate, the upper line of the gate coinciding with the surface of the water?

SOLUTION.

It is obvious that it will be somewhere on the line joining the middle points of the upper and lower edges of the gate.

Denote its distance from the upper edge, by z, the depth of the gate, by $2l$, and its mass, by M. The distance of the centre of gravity from the upper edge will be equal to l.

From Example 1 (Art. 132), replacing d by l, and reducing, we have, for the moment of inertia of the rectangle,

$$M\left(\frac{l^2}{3} + l^2\right) = M\tfrac{4}{3}l^2.$$

But the moment of the rectangle is equal to,

$$Ml;$$

hence, by division, we have,

$$z = \tfrac{4}{3}l = \tfrac{2}{3}(2l).$$

That is, the centre of pressure is at two-thirds of the distance from the upper to the lower edge of the gate.

2. Let it be required to find the pressure on a submerged rectangular flood-gate $ABCD$, the plane of the gate being vertical. Also, the distance of the centre of pressure below the surface of the water.

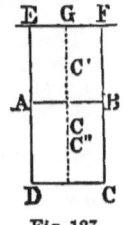

Fig. 137.

SOLUTION.

Let EF be the intersection of the plane with the surface of the water, and suppose the rectangle AC to be prolonged till it reaches EF. Let C, C', and C'', be the centres of pressure of the rectangles EC, EB, and AC respectively. Denote the distance GC'', by z, the distance ED, by a, and the distance EA, by a'. Denote the breadth of the gate, by b, and the weight, a unit of volume of the water, by w.

The pressure on EC will be equal to $\tfrac{1}{2}a^2bw$, and the pressure on EB will be equal to $\tfrac{1}{2}a'^2bw$; hence, the pressure on AC will be equal to

$$\tfrac{1}{2}bw(a^2 - a'^2);$$

which is the pressure required.

From the principle of moments, the moment of the pressure on AC, is equal to the moment of the pressure on EC, minus the moment of the pressure on EB. Hence, from the last problem,

$$\tfrac{1}{2}bw(a^2 - a'^2) \times z = \tfrac{1}{2}bwa^2 \times \tfrac{2}{3}a - \tfrac{1}{2}bwa'^2 \times \tfrac{2}{3}a',$$

$$\therefore \ z = \tfrac{2}{3}\frac{a^3 - a'^3}{a^2 - a'^2}$$

which is the required distance from the surface of the water.

3. Let it be required to find the pressure on a rectangular flood-gate, when both sides are pressed, the water being at different levels on the two sides. Also, to find the centre of pressure.

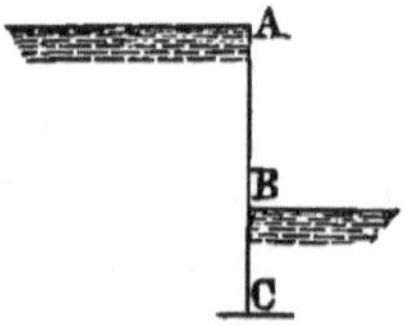

Fig. 188.

SOLUTION.

Denote the depth of water on one side by a, and on the other side, by a', the other elements being the same as before.

The total pressure will, as before, be equal to,

$$\tfrac{1}{2}bw(a^2 - a'^2).$$

Estimating z from C upwards,

$$z = \tfrac{1}{2}\frac{a^3 - a'^3}{a^2 - a'^2}. \quad Ans.$$

4. A sluice-gate, 10 feet square, is placed vertically, its upper edge coinciding with the surface of the water. What is the pressure on the upper and lower halves of the gate, respectively, the weight of a cubic foot of water being taken equal to 62½ lbs.? *Ans.* 7812.5 lbs., and 23437.5 lbs.

5. What must be the thickness of a rectangular dam of granite, that it may neither rotate about its outer angular

point nor slide along its base, the weight of a cubic foot of granite being 160 lbs., and the coefficient of friction between it and the soil being .6 ?

SOLUTION.

First, to find the thickness necessary to prevent rotation outwards. Denote the height of the wall, by h, and suppose the water to extend from the bottom to the top. Denote the thickness, by t, and the length of the wall, or dam, by l. The weight of the wall in pounds, will be equal to

$$lht \times 160;$$

and this being exerted through its centre of gravity, the moment of the weight with respect to the outer edge, as an axis, will be equal to

$$\tfrac{1}{2}t^2lh \times 160 = 80lht^2.$$

The pressure of the water against the inner face, in pounds, is equal to

$$\tfrac{1}{2}lh^2 \times 62.5 = lh^2 \times 31.25.$$

This pressure is applied at the centre of pressure, which is (Example 1) at a distance from the bottom of the wall equal to $\tfrac{1}{3}h$; hence, its moment with respect to the outer edge of the wall, is equal to

$$lh^3 \times 10.4166.$$

The pressure of the water tends to produce rotation outwards, and the weight of the wall acts to prevent this rotation. In order that these forces may be in equilibrium, their moments must be equal; or

$$80lht^2 = lh^3 \times 10.4166.$$

Whence, we find,

$$t = h\sqrt{.1302} = .36 \times h.$$

Next, to find the thickness necessary to prevent sliding along the base. The entire force of friction due to the weight of the wall, is equal to

$$160lht \times .6 = 96lht;$$

and in order that the wall may not slide, this must be equal to the pressure exerted horizontally against the wall. Hence,

$$96lht = 31.25lh^2.$$

Whence, we find,

$$t = .325h.$$

If the wall is made thick enough to prevent rotation, it will be secure against sliding.

6. What must be the thickness of a rectangular dam 15 feet high, the weight of the material being 140 lbs. to the cubic foot, that, when the water rises to the top, the structure may be just on the point of overturning?

Ans. 5.7 ft.

7. The staves of a cylindrical cistern filled with water, are held together by a single hoop. Where must the hoop be situated?

Ans. At a distance from the bottom equal to one-third of the height of the cistern.

8. Required the pressure of the sea on the cork of an empty bottle, when sunk to the depth of 600 feet, the diameter of the cork being $\frac{4}{5}$ of an inch, and a cubic foot of sea water being estimated to weigh 64 lbs.? *Ans.* 134 lbs.

Buoyant Effort of Fluids.

157. Let *A* represent any solid body suspended in a heavy fluid. Conceive this solid to be divided into vertical prisms, whose horizontal sections are infinitely small. Any one of these prisms will be pressed downward by a force equal to the weight of a column of fluid, whose base (Art. 155) is equal to the horizontal section of the filament, and whose altitude is the distance of its upper surface from the surface of the fluid; it will be pressed upward by a force equal to the weight of a column of fluid having the same base and an altitude equal to the distance of the lower base of the filament from the surface of the fluid. The resultant of these two pressures is a force exerted vertically upwards, and is equal to the weight of a column of fluid, equal in bulk to that of the filament and having its point of application at the centre of gravity of the volume of the filament. This being true for each filament of the body, and the lateral pressures being such as to destroy each other's effects, it follows, that the resultant of all the pressures upon the body will be a vertical force exerted upwards, whose intensity is equal to the weight of a portion of the fluid, whose volume is equal to that of the solid, and the point of application of which is the centre of gravity of the volume of the displaced fluid. This upward pressure is called the *buoyant effort* of the fluid, and its point of application is called *the centre of buoyancy*. The line of direction of the buoyant effort, in any position of the body, is called *a line of support*. That line of support which passes through the centre of gravity of a body, is called *the line of rest*.

Fig. 139.

Floating Bodies.

158. A body wholly or partially immersed in a heavy fluid, is urged downwards by its weight applied at its centre of gravity, and upwards, by the buoyant effort of the fluid applied at the centre of buoyancy.

The body can only be in equilibrium when the line through the centre of gravity of the body, and the centre of buoyancy, is vertical; in other words, when the line of rest is vertical. When the weight of the body exceeds the buoyant effort, the body will sink to the bottom; when they are just equal, it will remain in equilibrium, wherever placed in the fluid. When the buoyant effort is greater than the weight, it will rise to the surface, and after a few oscillations, will come to a state of rest, in such a position, that the weight of the displaced fluid is equal to that of the body, when it is said *to float*. The upper surface of the fluid is then called *the plane of floatation*, and its intersection with the surface of the body, *the line of floatation*.

If a floating body be slightly disturbed from its position of equilibrium, the centres of gravity and buoyancy will no longer be in the same vertical line. Let DE represent the plane of floatation, G the centre of gravity of the body (Fig. 141), GH its line of rest, and C the centre of buoyancy in the disturbed position of the body.

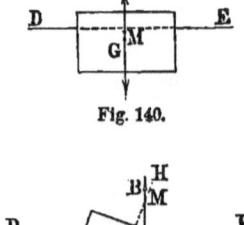

Fig. 140.

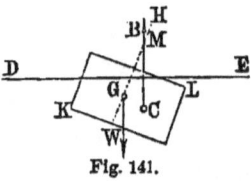

Fig. 141.

If the line of support CB, intersects the line of rest in M, above G, as in Fig. 141, the buoyant effort and the weight will conspire to restore the body to its position of equilibrium; in this case, the equilibrium must be *stable*.

If the point M falls below G, as in Fig. 142, the buoyant effort and the weight will conspire to overturn the body; in this case, the body must, before being disturbed, have been in a state of *unstable* equilibrium.

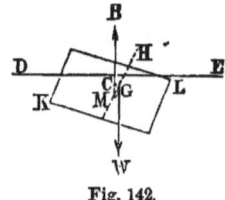

Fig. 142.

If the centre of buoyancy and centre of gravity are

always on the same vertical, the point
M will coincide with G (Fig. 143),
and the body will be in a state of
indifferent equilibrium. The limiting
position of the point M, or of the
intersection of the lines of rest and
of support, obtained by disturbing the
floating body through an infinitely small angle, is called the
metacentre of the body. Hence,

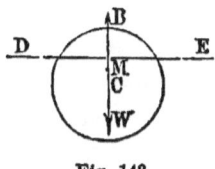

Fig. 143.

If the metacentre is above the centre of gravity of the body, it will be in a state of stable equilibrium, the line of rest being vertical; if it is below the centre of gravity, the body will be in unstable equilibrium; if the two points coincide, the body will be in indifferent equilibrium.

The stability of the floating body will be the greater, as the metacentre is higher above the centre of gravity. This condition is practically fulfilled in loading ships, or other floating bodies, by stowing the heavier objects nearest the bottom of the vessel.

Specific Gravity.

159. The specific gravity of a body is its *relative weight;* that is, it is the number of times the body is heavier than an equivalent volume of some other body taken as a standard.

The numerical value of the specific gravity of any body, is the quotient obtained by dividing the weight of any volume of the body by that of an equivalent volume of the standard.

For solids and liquids, water is generally taken as the standard, and, since this liquid is of different densities at different temperatures, it becomes necessary to assume also a standard temperature. Most writers have taken 60° Fahrenheit as this standard. Some, however, have taken 38°75 Fah., for the reason that experiment has shown that water has its maximum density at this temperature. We shall adopt the latter standard, remarking that specific gravities, determined at any temperature, may be readily reduced to what they would have been had they been determined at any other temperature.

The densities of pure water at different temperatures has been determined with great accuracy by experiment, and the results arranged in tables, the density at 38°75 being taken as 1.

Since the specific gravity of a body increases as the density of the standard diminishes, it will be a little less when referred to water at 38°75 than at any other temperature.

Let d and d' denote the densities of water at any two temperatures t and t'; let s and s' denote the specific gravities of the same body, referred to water at these temperatures; then,

$$s : s' : : d' : d, \quad \therefore s = \frac{s'd'}{d} \quad . \quad (146.)$$

This formula is applicable in any case where it is necessary to reduce the specific gravity taken at the temperature t' to what it would have been if taken at the temperature t. If $t = 38°75$, we have $d = 1$, and the formula becomes,

$$s = s'd' \quad . \quad . \quad . \quad . \quad . \quad (147.)$$

Hence, *to reduce the specific gravity taken at the temperature t', to the standard temperature, multiply it by the tabular density of water at the temperature t'.*

The specific gravity should also be corrected for expansion. This correction is made in a manner entirely similar to the last. Denote the volumes of the same body at the temperatures t and t', by v and v', and the apparent specific gravities, after the last correction, by S and S', then,

$$S : S' : : v' : v, \quad \therefore S = \frac{S'v'}{v} \quad (148.)$$

If t is the standard temperature, and v the unit of volume we have,

$$S = S' \times v' \quad \dots \quad (149.)$$

In what follows, we shall suppose that the specific gravities are taken at the standard temperature, in which case no correction will be necessary.

Gases are generally referred to atmospheric air as a standard, but, as air may be readily referred to water as a standard, we shall, for the purpose of simplification, suppose that the standard for *all* bodies is distilled water at 38°75 Fahrenheit.

Hydrostatic Balance.

160. This balance is similar to that described in Article 81, except the scale-pans have hooks attached to their lower surfaces for the purpose of suspending bodies. The suspension is effected by a fine platinum wire, or by some other material not acted upon by the liquids employed.

Fig. 144.

To determine the Specific Gravity of an Insoluble Body.

161. Attach the suspending wire to the first scale-pan, and after allowing it to sink in a vessel of water to a certain depth, counterpoise it by an equal weight, attached to the hook of the second scale-pan. Place the body in the first scale-pan, and counterpoise it by weights in the second pan. These weights will give the weight of the body in air. Next, attach the body to the suspending wire, and immerse it in the water. The buoyant effort of the water will be equal to the weight of a volume of water equivalent to that of the body (Art. 157); hence, the second pan will descend. Restore the equilibrium by weights placed in the first pan. These weights will give the weight of the displaced water

Divide the weight of the body in air by the weight just found, and the quotient will be the specific gravity sought. If the body will not sink in water, determine its weight in air as before; then attach to it a body so heavy, that the combination will sink; find, as before, the loss of weight of the combination, and also the loss of weight of the heavier body; take the latter from the former, and the difference will be the loss of weight of the lighter body; divide its weight in air by this weight, and the quotient will be the specific gravity sought.

If great accuracy is required, account must be taken of the buoyant effort of the air, which, when the body is very light, and of considerable dimensions, will render the apparent weight less than the true weight, or the weight in vacuum. Since the weights used in counterpoising are always very dense, and of small dimensions, the buoyant effort of the air upon them may always be neglected.

To determine the true weight of a body in vacuum: let w denote its weight in air, w' its weight in water, and W its weight in vacuum; then will $W - w$, and $W - w'$, denote its loss of weight in air and water; denote the specific gravity of air referred to water, by s. Since the losses of weight in air and water are proportional to their specific gravities, we have,

$$W - w : W - w' :: s : 1; \text{ or, } W - w = sW - sw',$$
$$\therefore W = \frac{w - sw'}{1 - s}.$$

This weight should be used, instead of the weight in air.

To determine the Specific Gravity of Liquids.

162. First Method.—Take a vial with a narrow neck, and weigh it; fill it with the liquid, and weigh again; empty out the liquid, and fill with water, and weigh again; deduct from the last two weights, respectively, the weight of the vial; these results will give the weights of equal volumes of the liquid and of water. Divide the former by the latter, and the quotient will be the specific gravity sought.

Second Method.—Take a heavy body, that will sink both in the liquid and in water, and which will not be acted upon by either; determine its loss of weight, as already explained, first in the liquid, then in water; divide the former by the latter, and the quotient will be the specific gravity sought. The reason is evident.

Third Method.—Let AB and CD represent two graduated glass tubes of half an inch in diameter, open at both ends. Let their upper ends communicate with the receiver of an air-pump, and their lower ends dip into two cisterns, one containing distilled water, and the other the liquid whose specific gravity is to be determined. Let the air be partially exhausted from the receiver by means of an air-pump; the liquids will rise in the tubes, but to different heights, these being inversely as the specific gravities of the liquids. If we divide the height of the column of water by that of the other liquid, the quotient will be the specific gravity sought. By creating different degrees of rarefaction, the columns will rise to different heights, but their ratios ought to be the same. We are thus enabled to make a series of observations, each corresponding to a different degree of rarefaction, from which a more accurate result can be had than from a single observation.

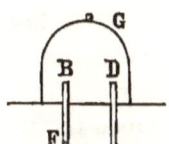

Fig. 145.

To determine the Specific Gravity of a Soluble Body.

163. Find its specific gravity by the method already given, with respect to some liquid in which it is not soluble, and find also the specific gravity of this liquid referred to water; take the product of these specific gravities, and it will be the specific gravity sought. For, if the body is m times heavier than an equivalent volume of the liquid used, and this is n times heavier than an equivalent volume of water, it follows that the body is mn times heavier than its volume of water, whence the rule.

The auxiliary liquid, in some cases, might be a saturated solution of the given body in water; the rule remains unchanged.

To determine the Specific Gravity of the Air.

164. Take a hollow globe, fitted with a stop-cock, to shut off communication with the external air, and, by means of the air-pump or condensing syringe, pump in as much air as is convenient, close the stop-cock, and weigh the globe thus filled. Provide a glass tube, graduated so as to show cubic inches and decimals of a cubic inch, and, having filled it with mercury, invert it over a mercury bath. Open the stopcock, and allow the compressed air to escape into the inverted tube, taking care to bring the tube into such a position that the mercury without the tube is at the same level as within. The reading on the tube will give the volume of the escaped air. Weigh the globe again, and subtract the weight thus found from the first weight; this difference will indicate the weight of the escaped air. Having reduced the measured volume of air to what it would have occupied at a standard temperature and barometric pressure, by means of rules yet to be deduced, compute the weight of an equivalent volume of water; divide the weight of the corrected volume of air by that of an equivalent volume of distilled water, and the quotient will be the specific gravity sought.

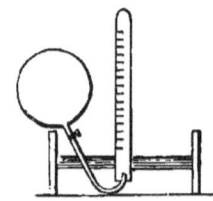

Fig. 146.

To determine the Specific Gravity of a Gas.

165. Take a glass globe of suitable dimensions, fitted with a stop-cock for shutting off communication with the atmosphere. Fill the globe with air, and determine the weight of the globe thus filled referred to a vacuum, as already explained. From the known volume of the globe and the specific gravity of air, the weight of the contained air can be computed; subtract this from the previous weight, and we shall have the true weight of the globe alone; determine in succession the weights of the globe filled with water and with the gas in vacuum, and from each subtract the weight of the globe; divide the latter result by the former; the quotient will be the specific gravity required.

Hydrometers.

166. A hydrometer is a floating body, used for the purpose of determining specific gravities. Its construction depends upon the principle of floatation. Hydrometers are of two kinds. 1. Those in which the submerged volume is constant. 2. Those in which the weight of the instrument remains constant.

Nicholson's Hydrometer.

167. This instrument consists of a hollow brass cylinder A, at the lower extremity of which is fastened a basket B, and at the upper extremity a wire, bearing a scale-pan C. At the bottom of the basket is a ball of glass E, containing mercury, the object of which is, to cause the instrument to float in an upright position. By means of this ballast, the instrument is adjusted so that a weight of 500 grains, placed in the pan C, will sink it in distilled water to a notch D, filed in the neck.

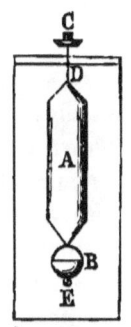

Fig. 147.

To determine the specific gravity of a solid which weighs less than 500 grains. Place the body in the pan C, and add weights till the instrument sinks, in distilled water, to the notch D. The added weights, substracted from 500 grains, will give the weight of the body in air. Place the body in the basket B, which generally has a reticulated cover, to prevent the body from floating away, and add other weights to the pan, until the instrument again sinks to the notch D. The weights last added give the weight of the water displaced by the body

Divide the first of these weights by the second, and the quotient will be the specific gravity required.

To find the specific gravity of a liquid. Having carefully weighed the instrument, place it in the liquid, and add weights to the scale-pan till it sinks to D. The weight of the instrument, plus the sum of the weights added, will be the weight of the liquid displaced by the instrument. Next, place the instrument in distilled water, and add weights till it sinks to D. The weight of the instrument, plus the added weights, gives the weight of the displaced water. Divide the first result by the second, and the quotient will be the specific gravity required. The reason for this rule is evident.

A modification of this instrument, in which the basket B, is omitted, is sometimes constructed for determining specific gravities of liquids only. This kind of hydrometer is generally made of glass, that it may not be acted upon chemically, by the liquids into which it is plunged. The hydrometer just described, is generally known as *Fahrenheit's hydrometer*, or *Fahrenheit's areometer*.

Scale Areometer.

168. The scale areometer is a hydrometer whose weight remains constant; the specific gravity of a liquid is made known by the depth to which it sinks in it. The instrument consists of a hollow glass cylinder A, with a stem C, of uniform diameter. At the bottom of the cylinder is a bulb B, containing mercury, to make the instrument float upright. By introducing a suitable quantity of mercury, the instrument may be adjusted so as to float at any desired point of the stem. When it is designed to determine the specific gravities of liquids, both heavier and lighter than water, it is ballasted so that in distilled water, it will sink to the middle of the stem. This point is marked on the stem with a file, and since the specific gravity of water is 1, it is numbered 1 on the scale. A liquid is then formed by dissolving common salt in water whose specific gravity is

Fig. 143.

1.1, and the instrument is allowed to float freely in it; the point E, to which it then sinks, is marked on the stem, and the intermediate part of the scale, HE, is divided into 10 equal parts, and the graduation continued above and below throughout the stem. The scale thus constructed is marked on a piece of paper placed within the hollow stem. To use this hydrometer, we have simply to put it into the liquid and allow it to come to rest; the division of the scale which corresponds to the surface of floatation, makes known the specific gravity of the liquid. The hypothesis on which this instrument is graduated, is, that the increments of specific gravity are proportional to the increments of the submerged portion of the stem. This hypothesis is only approximately true, but it approaches more nearly to the truth as the diameter of the stem diminishes.

When it is only desired to use the instrument for liquids heavier than water, the instrument is ballasted so that the division 1 shall come near the top of the stem. If it is to be used for liquids lighter than water, it is ballasted so that the division 1 shall fall near the bottom of the stem. In this case we determine the point 0.9 by using a mixture of alcohol and water, the *principle* of graduation being the same as in the first instance.

Volumeter.

169. The volumeter is a modification of the scale areometer, differing from it only in the method of graduation. The graduation is effected as follows: The instrument is placed in distilled water, and allowed to come to a state of rest, and the point on the stem where the surface cuts it, is marked with a file. The submerged volume is then accurately determined, and the stem is graduated in such a manner that each division indicates a volume equal to a hundredth part of the volume originally submerged. The divisions are then numbered from the first mark in both directions, as indicated in the figure. To use the instrument, place it in the iquid, and note the division to which it sinks;

Fig. 149.

divide 100 by the number indicated, and the quotient will be the specific gravity sought. The principle employed is, that the specific gravities of liquids are inversely as the volumes of equal weights. Suppose that the instrument indicates x parts; then the weight of the instrument displaces x parts of the liquid, whilst it displaces 100 parts of water. Denoting the specific gravity of the liquid by S, and that of water by 1, we have,

$$S : 1 :: 100 : x, \quad \therefore S = \frac{100}{x}.$$

A table may be computed to save the necessity of performing the division.

Densimeter.

170. The densimeter is a modification of the volumeter, and admits of use when only a small portion of the liquid can be had, as is often the case in examining animal secretions, such as bile, chyle, &c. The construction of the densimeter differs from that of the volumeter, last described, in having a small cup at the upper extremity of the stem, destined to receive the fluid whose specific gravity is to be determined.

The instrument is ballasted so that when the cup is empty, the densimeter will sink in distilled water to a point B, near the bottom of the stem. This point is the 0 of the instrument. The cup is then filled with distilled water, and the point C, to which it sinks, is marked; the space BC, is divided into any number of equal parts, say 10, and the graduation is continued to the top of the tube.

Fig. 150.

To use the instrument, place it in distilled water, and fill the cup with the liquid in question, and note the division to which it sinks. Divide 10 by the number of this division, and the quotient will be the specific gravity required. The principle of the densimeter is the same as that of the volumeter.

MECHANICS OF LIQUIDS. 261

Centesimal Alcoholometer of Gay Lussac.

171. This instrument is the same in construction as the scale areometer; the graduation is, however, made on a different principle. Its object is, to determine the percentage of alcohol in a mixture of alcohol and water. The graduation is made as follows: the instrument is first placed in absolute alcohol, and ballasted so that it will sink nearly to the top of the stem. This point is marked 100. Next, a mixture of 95 parts of alcohol and 5 of water, is made, and the point to which the instrument sinks, is marked 95. The intermediate space is divided into 5 equal parts. Next, a mixture of 90 parts of alcohol and 10 of water is made; the point to which the instrument sinks, is marked 90, and the space between this and 95, is divided into 5 equal parts. In this manner, the entire stem is graduated by successive operations. The spaces on the scale are not equal at different points, but, for a space of five parts, they may be regarded as equal, without sensible error.

To use the instrument, place it in the mixture of alcohol and water, and read the division to which it sinks; this will indicate the percentage of alcohol in the mixture.

In all of the instruments, the temperature has to be taken into account; this is usually effected by means of corrections, which are tabulated to accompany the different instruments.

On the principle of the alcoholometer, are constructed a great variety of areometers, for the purpose of determining the degrees of saturation of wines, syrups, and other liquids employed in the arts.

In some nicely constructed hydrometers, the mercury used as ballast serves also to fill the bulb of a delicate thermometer, whose stem rises into the cylinder of the instrument, and thus enables us to note the temperature of the fluid in which it is immersed.

EXAMPLES.

1. A cubic foot of water weighs 1000 ounces. Required

the weight of a cubical block of stone, one of whose edges is 4 feet, its specific gravity being 2.5. *Ans.* 10000 lbs.

2. Required the number of cubic feet in a body whose weight is 1000 lbs., its specific gravity being 1.25.
Ans. 12.8.

3. Two lumps of metal weigh respectively 3 lbs., and 1 lb., and their specific gravities are 5 and 9. What will be the specific gravity of an alloy formed by melting them together, supposing no contraction of volume to take place.
Ans. 5.625.

4. A body weighing 20 grains has a specific gravity of 2.5. Required its loss of weight in water. *Ans.* 8 grains.

5. A body weighs 25 grains in water, and 40 grains in a liquid whose specific gravity is .7. What is the weight of the body in vacuum? *Ans.* 75 grains.

6. A Nicholson's hydrometer weighs 250 grains, and it requires an additional weight of 336 grains to sink it to the notch in the stem, in a mixture of alcohol and water. What is the specific gravity of the mixture? *Ans.* .781.

7. A block of wood is found to sink in distilled water till $\frac{7}{8}$ of its volume is submerged. What is its specific gravity?
Ans. .875.

8. The weight of a piece of cork in air, is $\frac{3}{4}$ oz.; the weight of a piece of lead in water, is $6\frac{4}{5}$ oz.; the weight of the cork and lead together in water, is $4\frac{7}{100}$ oz. What is the specific gravity of the cork? *Ans.* 0.24.

9. A solid, whose weight is 250 grains, weighs in water, 147 grains, and, in another fluid, 120 grains. What is the specific gravity of the latter fluid? *Ans.* 1.26°.

10. A solid weighs 60 grains in air, 40 in water, and 30 in an acid. What is the specific gravity of the acid?
Ans. 1.5.

MECHANICS OF LIQUIDS. 263

The following table of the specific gravity of some of the most important solid and fluid bodies, is compiled from a table given in the *Ordnance Manual.*

TABLE OF SPECIFIC GRAVITIES OF SOLIDS AND LIQUIDS.

SOLIDS.	SPEC. GRAV.	SOLIDS.	SPEC. GRAV.
Antimony, cast	6.712	Limestone	3.180
Brass, cast	8.396	Marble, common	2.686
Copper, cast	8.788	Salt, common	2.130
Gold, hammered	19.361	Sand	1.800
Iron, bar	7.788	Slate	2.672
Iron, cast	7.207	Stone, common	2.520
Lead, cast	11.352	Tallow	0.945
Mercury at 32° F	13.598	Boxwood	0.912
" at 60°	13.580	Cedar	0.596
Platina, rolled	22.069	Cherry	0.715
" hammered	20.337	Lignum vitæ	1.333
Silver, hammered	10.511	Mahogany	0.854
Tin, cast	7.291	Oak, heart	1.170
Zinc, cast	6.861	Pine, yellow	0.660
Bricks	1.900	Nitric acid	1.217
Chalk	2.784	Sulphuric acid	1.841
Coal, bituminous	1.270	Alcohol, absolute	0.792
Diamond	3.521	Ether, sulphuric	0.715
Earth, common	1.500	Sea water	1.026
Gypsum	2.168	Olive oil	0.915
Ivory	1.822	Oil of Turpentine	0.870

Thermometer.

172. A thermometer is an instrument used for measuring the temperatures of bodies. It is found, by observation, that almost all bodies expand when heated, and contract when cooled, so that, other things being equal, they always occupy the same volumes at the same temperatures. It is also found that different bodies expand and contract in a different ratio for the same increments of temperature. As a general rule, liquids expand much more rapidly than solids, and gases much more rapidly than liquids. The construction of the thermometer depends upon this principle of unequal expansibility of different bodies. A great variety of combinations have been used in the construction of thermometers, only one of which, the common *mercurial thermometer*, will be described.

The mercurial thermometer consists of a cylindrical or spherical bulb *A*, at the upper extremity of which, is a narrow tube of uniform bore, hermetically sealed at its upper end. The bulb and tube are nearly filled with mercury, and the whole is attached to a frame, on which is a scale for determining the temperature, which is indicated by the rise and fall of the mercury in the tube.

Fig. 151.

The tube should be of uniform bore throughout, and, when this is the case, it is found that the relative expansion of the mercury and glass is very nearly uniform for constant increments of temperature. A thermometer may be constructed and graduated as follows: A tube of uniform bore is selected, and upon one extremity a bulb is blown, which may be cylindrical or spherical; the former shape is, on many accounts, the preferable one. At the other extremity, a conical-shaped funnel is blown open at the top. The funnel is filled with mercury, which should be of the purest quality, and the whole being held vertical, the heat of a spirit-lamp is applied to the bulb, which expanding the air contained in it, forces a portion in bubbles up through the mercury in the funnel. The instrument is next allowed to cool, when a portion of mercury is forced down the capillary tube into the bulb. By a repetition of this process, the entire bulb may be filled with mercury, as well as the tube itself. Heat is then applied to the bulb, until the mercury is made to boil; and, on being cooled down to a little above the highest temperature which it is desired to measure, the top of the tube is melted off by means of a jet of flame, urged by a blow-pipe, and the whole is hermetically sealed. The instrument, thus prepared, is attached to a frame, and graduated as follows:

The instrument is plunged into a bath of melting ice, and, after being allowed to remain a sufficient time for the parts of the instrument to take the uniform temperature of the melting ice, the height of the mercury in the tube is marked on the scale. This gives the *freezing point* of the scale. The instrument is next plunged into a bath of boiling water, and allowed to remain long enough for all of the parts to acquire the temperature of the water and steam. The height of the mercury is then marked on the scale. This gives the *boiling point* of the scale. The freezing and boiling points having been determined, the intermediate space is divided into a certain number of equal parts, according to the scale adopted, and the graduation is then continued, both upwards and downwards, to any desired extent. Three principal scales are used. FAHRENHEIT'S *scale*, in which the space between the freezing and boiling point is divided into 180 equal parts, called degrees, the freezing point being marked 32°, and the boiling point 212°. In this scale, the 0 point is 32 degrees below the freezing point. *The Centigrade scale*, in which the space between the fixed points is divided into 100 equal parts, called degrees. The 0 of this scale is at the freezing point. REAUMUR'S *scale*, in which the same space is divided into 80 equal parts, called degrees. The 0 of this scale also is at the freezing point.

If we denote the number of degrees on the Fahrenheit, Centigrade, and Reaumur scales, by F, C, and R respectively, the following formula will enable us to pass from any one of these scales to any other:

$$\tfrac{1}{9}(F° - 32) = \tfrac{1}{5}C° = \tfrac{1}{4}R°.$$

The scale most in use in this country is FAHRENHEIT'S. The other two are much used in Europe, particularly the Centigrade scale.

Velocity of a liquid flowing through a small orifice.

173. Let ABD represent a vessel, having a very small orifice at its bottom, and filled with any liquid.

Denote the area of the orifice, by a, and its depth below the upper surface, by h. Let D represent an infinitely small layer of the liquid situated at the orifice, and denote its height, by h'. This layer is (Art. 155) urged downwards by a force equal to the weight of a column of the liquid whose base is equal to the orifice, and whose height is h; denoting this pressure, by p, and the weight of a unit of volume of the liquid, by w, we shall have,

Fig. 152.

$$p = wah.$$

If the element is pressed downwards by its own weight alone, this pressure being denoted by p', we have,

$$p' = wah'.$$

Dividing the former equation by the latter, member by member, we have,

$$\frac{p}{p'} = \frac{h}{h'};$$

that is, *the pressures are to each other as the heights h and h'.*

Were the element to fall through the small height h', under the action of the pressure p', or its own weight, the velocity generated would (Art. 115) be given by the equation,

$$v' = \sqrt{2gh'}.$$

Denoting the velocity actually generated whilst the element is falling throught the height h', by v, and recollecting that the velocities generated in falling through a given height, are to each other as the square roots of the pressures, we shall have,

$$v : v' :: \sqrt{p} : \sqrt{p'}, \quad \therefore v = v'\sqrt{\frac{p}{p'}}.$$

Substituting for v' its value, just deduced, and for $\dfrac{p}{p'}$ its value, $\dfrac{h}{h'}$, we have

$$v = \sqrt{2gh} \quad \ldots \ldots \quad (150.)$$

Hence, we conclude that *a liquid will issue from a very small orifice at the bottom of the containing vessel, with a velocity equal to that acquired by a heavy body in falling freely through a height equal to the depth of the orifice below the surface of the fluid.*

We have seen that the pressure due to the weight of a fluid upon any point of the surface of the containing vessel, is normal to the surface, and is always proportional to the depth of the point below the level of the free surface. Hence, if the side of a vessel be thin, so as not to affect the flow of the liquid, and an orifice be made at any point, the liquid will flow out in a jet, normal to the surface at the opening, and with a velocity due to a height equal to that of the orifice from the free surface of the fluid.

If the orifice is on the vertical side of a vessel, the initial direction of the jet will be horizontal; if it be made at a point where the tangent plane is oblique to the horizon, the initial direction of the jet will be oblique; if the opening is made on the upper side of a portion of a vessel where the tangent is horizontal, the jet will be directed upwards, and will rise to a height due to the velocity; that is, to the height of the upper surface of the fluid. This

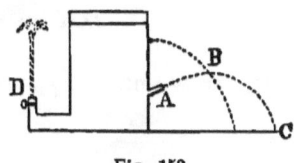

Fig. 158.

can be illustrated experimentally, by introducing a tube near the bottom of a vessel of water, and bending its outer extremity upwards, when the fluid will be observed to rise to the level of the upper surface of the water in the vessel.

Spouting of Liquids on a Horizontal Plane.

174. Let KL represent a vessel filled with water. Let D represent an orifice in its vertical side, and DE the path described by the spouting fluid. We may regard each drop of water as it issues from the orifice, as a projectile shot forth horizontally, and then acted upon by the force of gravity. Its path will, therefore, be a parabola, and the circumstances of its motion will be made known by a discussion of Equations (115) and (120).

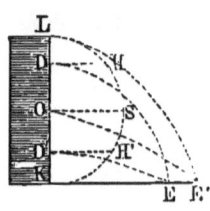

Fig. 154.

Denote the distance DK, by h', and the distance DL, by h. We have, from Equation (120), by making y equal to h', and $x = KE$,

$$KE = \sqrt{\frac{2v^2 h'}{g}}$$

But we have found that $v = \sqrt{2gh}$; hence, by substitution, we have,

$$KE = 2\sqrt{hh'}.$$

If we describe a semicircle on KL, as a diameter, and through D draw an ordinate DH, we shall have, from a well-known property of the circle,

$$DH = \sqrt{DK \cdot DL} = \sqrt{hh'}.$$

Hence we have, by substitution,

$$KE = 2\overline{DH}.$$

Since there are two points on KL at which the ordinates are equal, it follows that there are two orifices through which the fluid will spout to the same distance on the horizontal plane; one of these will be as far above the centre O, as the other is below it.

If the orifice be at O, midway between K and L, the ordinate OS will be the greatest possible, and the range KE' will be a maximum. The range in this case will be equal to the diameter of the circle LHK, or to the distance from the level of the water in the vessel to the horizontal plane.

If a semi-parabola LE' be described, having its axis vertical, its vertex at L, and focus at K, then may every point P, within the curve, be reached by two separate jets issuing from the side of the vessel; every point on the curve can be reached by one, and only one; whilst points lying without the curve cannot be reached by any jet whatever.

If the jet is directed obliquely upwards by a short pipe A (Fig. 153), the path described by each particle will still be the arc of a parabola ABC. Since each particle of the liquid may be regarded as a body projected obliquely upward, the nature of the path and the circumstances of the motion will be given by Equation (115).

In like manner, a discussion of the same equation will make known the nature of the path and the circumstances of motion, when the jet is directed obliquely downwards by means of a short tube.

Modifications due to extraneous pressure.

175. If we suppose the upper surface of the liquid, in any of the preceding cases, to be pressed by any force, as when it is urged downwards by a piston, we may denote the height of a column of fluid whose weight is equal to the extraneous pressure, by h'. The velocity of efflux will then be given by the equation,

$$v = \sqrt{2g(h + h')}.$$

The pressure of the atmosphere acts equally on the upper surface and the surface of the opening; hence, in ordinary cases, it may be neglected; but were the water to flow into a vacuum, or into rarefied air, the pressure must be taken into account, and this may be done by means of the formula just given.

Should the flow take place into condensed air, or into any medium which opposes a greater resistance than the atmospheric pressure, the extraneous pressure would act upwards, h' would be negative, and the preceding formula would become,

$$v = \sqrt{2g(h - h')}.$$

Coefficients of Efflux and Velocity.

176. When a vessel empties itself through a small orifice at its bottom, it is observed that the particles of fluid near the top descend in vertical lines; when they approach the bottom they incline towards the orifice, the converging lines of fluid particles tending to cross each other as they emerge from the vessel. The result is, that the stream grows narrower, after leaving the vessel, until it reaches a point at a distance from the vessel equal to about the radius of the orifice, when the contraction becomes a minimum, and below that point the vein again spreads out. This phenomenon is called the contraction of the vein. The cross section at the most contracted part of the vein, is not far from $\frac{64}{100}$ of the area of the orifice, when the vessel is very thin. If we denote the area of the orifice, by a, and the area of the least cross section of the vein, by a', we shall have,

$$a' = ka,$$

in which k is a number to be determined by experiment. This number is called the *coefficient of contraction*.

To find the quantity of water discharged through an orifice at the bottom of the containing vessel, in a second, we have only to multiply the area of the smallest cross section of the vein, by the velocity. Denoting the quantity discharged in one second, by Q', we shall have,

$$Q' = ka\sqrt{2gh}.$$

This formula is only true on the supposition that the actual velocity is equal to the theoretical velocity, which is not the case, as has been shown by experiment. The theoretical velocity has been shown to be equal to $\sqrt{2gh}$, and if we denote the actual velocity, by v', we shall have,

$$v' = l\sqrt{2gh},$$

in which l is to be determined by experiment; this value of l is slightly less than 1, and is called the *coefficient of velocity*. In order to get the actual discharge, we must replace $\sqrt{2gh}$ by $l\sqrt{2gh}$, in the preceding equation. Doing so, and denoting the actual discharge per second, by Q, we have,

$$Q = kla\sqrt{2gh}.$$

The product kl, is called the *coefficient of efflux*. It has been shown by experiment, that this coefficient for orifices in thin plates, is not quite constant. It decreases slightly, as the area of the orifice and the velocity are increased; and it is further found to be greater for circular orifices than for those of any other shape.

If we denote the coefficient of efflux, by m, we have,

$$Q = ma\sqrt{2gh}.$$

In this equation, h is called the *head of water*. Hence, we may define the *head of water* to be the distance from the orifice to the plane of the upper surface of the fluid.

The mean value of m corresponding to orifices of from $\frac{1}{2}$ to 6 inches in diameter, with from 4 to 20 feet head of water, has been found to be about .615. If we take the value of $k = .64$, we shall have,

$$l = \frac{m}{k} = \frac{.615}{.640} = .96.$$

That is, the actual velocity is only $\frac{96}{100}$ of the theoretical velocity. This diminution is due to friction, viscosity, &c.

Efflux through Short Tubes.

177. It is found that the discharge from a given orifice is increased, when the thickness of the plate through which the flow takes place, is increased; also, when a short tube is introduced.

When a tube AB, is employed which is not more than four times as long as the diameter of the orifice, the value of m becomes, on an average, equal to .813; that is, the discharge per second is 1.325 times greater when the tube is used, than without it. In using the cylindrical tube, the contraction takes place at the outlet of the vessel, and not at the outlet of the tube.

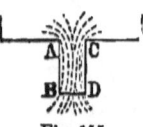

Fig. 155.

Compound mouth-pieces are sometimes used formed of two conic frustrums, as shown in the figure, having the form of the vein. It has been shown by Etelwein, that the most effective tubes of this form should have the diameter of the cross section CD, equal to .833 of the diameter AB. The angle made by the sides CE and DF, should be about 5° 9′, and the length of this portion should be three times that of the other.

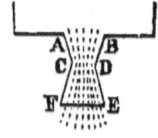

Fig. 156.

EXAMPLES.

1. With what theoretical velocity will water issue from a small orifice $16\frac{1}{13}$ feet below the surface of the fluid?

Ans. $32\frac{1}{6}$ ft.

2. If the area of the orifice, in the last example, is $\frac{1}{10}$ of a square foot, and the coefficient of efflux .615, how many cubic feet of water will be discharged per minute?

Ans. 118.695 ft.

3. A vessel, constantly filled with water, is 4 feet high, with a cross-section of one square foot; an orifice in the bottom has an area of one square inch. In what time will three-fourths of the water be drawn off, the coefficient of efflux being .6? *Ans.* $\frac{3}{4}$ minute, nearly.

4. A vessel is kept constantly full of water. How many cubic feet of water will be discharged per minute from an orifice 9 feet below the upper surface, having an area of 1 square inch, the coefficient of efflux being .6?

Ans. 6 cubic feet, about.

5. In the last example, what will be the discharge per minute, if we suppose each square foot of the upper surface to be pressed by a force of 645 lbs.?

Ans. $8\frac{3}{4}$ cubic feet, about.

6. The head of water is 16 feet, and the orifice is $\frac{1}{100}$ of a square foot. What quantity of water will be discharged per second, when the orifice is through a thin plate?

SOLUTION.

In this case, we have,

$$Q = .615 \times .01 \sqrt{2 \times 32\tfrac{1}{6} \times 16} = .197 \text{ cubic feet.}$$

When a short cylindrical tube is used, we have,

$$Q = .197 \times 1.325 = .261 \text{ cubic feet.}$$

In ETELWEIN's compound mouthpiece, if we take the smallest cross-section as the orifice, and denote it by a, it is found that the discharge is $2\frac{1}{2}$ times that through an orifice of the same size in a thin plate. In this case, we have, supposing $a = \frac{1}{100}$ of a square foot,

$$Q = .197 \times 2\frac{1}{2} = .49 \text{ cubic feet.}$$

12*

Motion of water in open channels.

178. When water flows through an open channel, as in a river, canal, or open aqueduct, the form of the channel being always the same, and the supply of water being constant, it is a matter of observation that the flow becomes uniform; that is, the quantity of water that flows through any cross-section, in a given time, is constant. On account of adhesion, friction, &c., the particles of water next the sides and bottom of the channel have their motion retarded. This retardation is imparted to the next layer of particles, but in a less degree, and so on, till a line of particles is reached whose velocity is greater than that of any other filament. This line, or filament of particles, is called the axis of the stream. In the case of cylindrical pipes, the axis coincides sensibly with the axis of the pipe; in straight, open channels, it coincides with that line of the upper surface which is midway between the sides.

A section at right-angles to the axis is called a *cross-section*, and, from what has been shown, the velocities of the fluid particles will be different at different points of the same cross-section. The *mean velocity* corresponding to **any** cross-section, is the average velocity of the particles at **every** point of that section. The mean velocity may be found by dividing the volume which flows through the section in one second, by the area of the cross-section. Since the same volume flows through each cross-section per second, after the flow has become uniform, it follows that, in channels of varying width, the mean velocity, at any section, will be inversely as the area of the section.

The intersection of the plane of cross-section with the sides and bottom of the channel, is called the *perimeter* of the section. In the case of a pipe which is constantly filled, the perimeter is the entire line of intersection of the plane of cross-section, with the interior surface of the pipe.

The mean velocity of water in an open channel depends, in the first place, upon its inclination to the horizon. As the inclination becomes greater, the component of gravity in the direction of the channel increases, and, consequently, the velocity becomes greater. Denoting the inclination by I, and resolving the force of gravity into two components, one at right angles to the upper surface, and the other parallel to it, we shall have for the latter component,

$$g\sin I.$$

This is the only force that acts to increase the velocity. The velocity will be diminished by friction, adhesion, &c. The total effect of these resistances will depend upon the ratio of the perimeter to the area of the cross section, and also upon the velocity. The cross-section being the same, the resistances will increase as the perimeter increases; consequently, for the same cross-section, the resistance of friction will be the least possible when the perimeter is least possible. The retardation of the flow will also diminish as the area of the cross-section is increased, other things remaining unchanged.

If we denote the area of the cross-section by a, the perimeter, by P, and the velocity, by v, we shall have,

$$\frac{ag\sin I}{P} = f(v),$$

in which f denotes some function of v.

Since the inclination is very small in all practical cases, we may place the inclination itself for the sine of the inclination, and doing so, it has been shown by PRONY, that the function of v may be expressed by two terms, one of which is of the first, and the other of the second degree, with respect to v; or,

$$\frac{gaI}{P} = mv + nv^2.$$

Denoting $\frac{a}{P}$ by R, $\frac{m}{g}$ by k, and $\frac{n}{g}$ by l, we have, finally,

$$kv + lv^2 = RI.$$

in which k and l are constants, to be determined by experiment. According to ETELWEIN, we have,

$$k = .0000242651, \text{ and } l = .0001114155.$$

Substituting these values, and solving with respect to v, we have,

$$v = -0.1088941604 + \sqrt{.0118580490 + 8975.414285 RI},$$

from which the velocity can be found when R and I are known. The values of k and l, and consequently that of v, were found by PRONY to be somewhat different from those given above. Those of ETELWEIN are selected for the reason that they were based upon a much larger number of experiments than those of PRONY.

Having the mean velocity and the area of the cross-section, the quantity of water delivered in any time can be computed. Denoting the quantity delivered in n seconds, by Q, and retaining the preceding notation, we have,

$$Q = nav.$$

The quantity of water to be delivered is generally one of the data in all practical problems involving the distribution of water. The difference of level of the point of supply and delivery is also known. The preceding principles enable us to give such a form to the cross-section of the canal, or aqueduct, as will ensure the requisite supply.

Were it required to apply the results just deduced, to the case of irregular channels, or to those in which there were many curves, a considerable modification would be required. The theory of these modifications does not come within the limits assigned to this treatise. For a complete discussion of the whole subject of hydraulics in a popular form, the reader is referred to the *Traité d'Hydraulique* D'AUBISSON

Motion of water in pipes.

179. The circumstances of the motion of water in pipes, are closely analagous to those of its motion in open channels. The forces which tend to impart motion are dependent upon the weight of the water in the pipe, and upon the height of the water in the upper reservoir. Those which tend to prevent motion depend upon the depth of water in the lower reservoir, friction in the pipe, adhesion, and shocks arising from irregularities in the bore of the pipe. The retardation due to shocks will, for the present, be neglected.

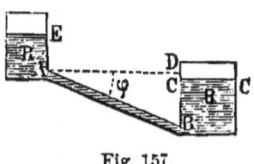

Fig. 157.

Let AB represent a straight cylindrical pipe, connecting two reservoirs R and R'. Suppose the water to maintain its level at E, in the upper, and at C, in the lower reservoir. Denote AE, by h, and BC, by h'. Denote the length of the pipe, by l, its circumference, by c, its cross-section, by a, its inclination, by φ, and the weight of a unit of volume of water, by w.

Experience shows that, under the circumstances above indicated, the flow soon becomes uniform. We may then regard the entire mass of fluid in the pipe as a coherent solid, moving with a mean uniform velocity down the inclined plane AB.

The weight of the water in the pipe will be equal to wal. If we resolve this weight into two components, one perpendicular to, and the other coinciding with the axis of the tube, we shall have for the latter component, $wal \sin\varphi$. But $l \sin\varphi$ is equal to DB. Denoting this distance by h'', we shall have for the pressure in the direction of the axis, due to the weight of the water in the pipe, the expression wah''. This pressure acts from A towards B. The pressure due to the weight of the water in R, and acting in the same direction, is wah.

The forces acting from B towards A, are, *first*, that due to the weight of the water in R', which is equal to wah'; and, *secondly*, the resistance due to friction and adhesion. This resistance depends upon the length of the pipe, its circumference and the velocity. It has been shown, by experiment, that this force may be expressed by the term,

$$cl(kv + k'v^2).$$

Since the velocity has been supposed uniform, the forces acting in the direction of the axis, must be in equilibrium. Hence,
$$wah + wah'' = wah' + cl(kv + k'v^2);$$

whence, by reduction,

$$\frac{k}{w}v + \frac{k'}{w}v^2 = \frac{a}{c}\left(\frac{h + h'' - h'}{l}\right).$$

The factor $\frac{a}{c}$ is equal to one-fourth of the diameter of the pipe. Denoting this by d, we shall have, $\frac{a}{c} = \tfrac{1}{4}d$; denoting $\frac{k}{w}$ by m, $\frac{k'}{w}$ by n, and $\frac{h+h''-h'}{l}$ by s, we have,

$$mv + nv^2 = \tfrac{1}{4}ds.$$

The values of m and n, as determined experimentally by Prony, are,
$$m = 0.00017, \quad \text{and} \quad n = 0.000106.$$

Hence, by substitution,

$$.00017v + .000106v^2 = \tfrac{1}{4}ds.$$

If v is not very small, the first term may be neglected, which will give,
$$v = 48.56\sqrt{ds}.$$

If we denote the quantity of water delivered in n seconds, by Q, we shall have,

$$Q = nav = 48.56na\sqrt{ds}.$$

The velocity will be greatly diminished, if the tube is curved to any considerable extent, or if its diameter is not uniform throughout. It is not intended to enter into a discussion of these cases; their complete development would require more space than has been allotted to this branch of Mechanics.

General Remarks on the distribution and flow of water in pipes.

180. Whenever an obstacle occurs in the course of an open channel or pipe, a change of velocity must take place. In passing the obstacle, the velocity of the water will increase, and then, impinging upon that which has already passed, a shock will take place. This shock consumes a certain amount of living force, and thus diminishes the velocity of the stream. All obstacles should be avoided; or, if any are unavoidable, the stream should be diminished, and again enlarged gradually, so as to avoid, as much as possible, the necessary shock incident to sudden changes of velocity.

For a like reason, when a branch enters the main channel, it should be made to enter as nearly in the direction of the current as possible.

All changes of direction give rise to mutual impacts amongst the particles, and the more, as the change is more abrupt. Hence, when a change of direction is necessary, the straight branches should be made tangential to the curved portion.

The entrance to, and outlet from a pipe or channel, should be enlarged, in order to diminish, as much as possible, the coefficients of ingress and egress.

When a pipe passes over uneven ground, sometimes ascending, and sometimes descending, there is a tendency to a collection of bubbles of air, at the highest points, which may finally come to act as an impeding cause to the flow. There should, therefore, be suitable pipes inserted at the highest points, to permit the confined air to escape.

Finally, attention should be given to the form of the cross-section of the channel. If the channel is a pipe, it should be made cylindrical. If it is a canal or open aqueduct, that form should be given to the perimeter which would give the greatest cross-section, and, at the same time, conform to the necessary conditions of the structure. The perimeter in open channels is generally trapezoidal, from the necessity of the case; and it should be remembered, that the nearer the form approaches a semi-circle, the greater will be the flow.

Capillary Phenomena.

181. When a liquid is in equilibrium, under the action of its own weight, it has been shown that its upper surface is level. It is observed, however, in the neighborhood of solid bodies, such as the walls of a containing vessel, that the surface is sometimes elevated, and sometimes depressed, according to the nature of the liquid and solid in contact. These elevations and depressions result from the action of molecular forces, exerted between the particles of the liquid and solid which are in contact; from the fact that they are more apparent in the case of small tubes, of the diameter of a hair, these phenomena have been called *capillary phenomena*, and the forces giving rise to them, *capillary forces*.

These forces only produce sensible effects at extremely small distances. CLAIRAUT has shown, that when the intensity of the force of attraction of the particles of the solid for those of the liquid, exceeds one-half that of the particles of the liquid for each other, the liquid will be *elevated* about the solid; when less, it will be *depressed*; when equal, it will *neither be elevated nor depressed*. In the first case, the resultant of the capillary forces is a force of *capillary attraction;* in the second case, it is a force of *capillary repulsion;* and in the third case, the capillary forces are in equilibrium.

The following are some of the observed effects of capillary

action: When a solid is plunged into a liquid which is capable of moistening it, as when wood or glass is plunged into water, the surface of the liquid is heaped up about the solid, taking a concave form, as shown in Fig. 158.

When a solid is plunged into a liquid which is not capable of moistening it, as when glass is plunged into mercury, the surface of the liquid is depressed about the solid, taking a convex form, as shown in Fig. 159.

Fig. 158.

Fig. 159.

The surface of the liquid in the neighborhood of the bounding surfaces of the containing vessel takes the form of concavity or convexity, according as the material of the vessel is capable of being moistened, or not, by the liquid.

These phenomena become more apparent when, instead of a solid body, we plunge a tube into a liquid, according as the material of the tube is, or is not, capable of being moistened by the liquid, the liquid will rise in the tube or be depressed in it. When the liquid rises in the tube, its upper surface takes a concave shape; when it is depressed, it takes a convex form. The elevations or depressions increase as the diameter of the tube diminishes.

Elevation and Depression between plates.

182. If two plates of any substance are placed parallel to each other, it is found that the laws of ascent and descent of the liquid into which they are plunged, are essentially the same as for tubes. For example: if two plates of glass parallel to each other, and pretty close together, are plunged into water, it is found that the water will rise between them to a height which is inversely proportional to their distance apart; and further, that this height is equal to half the height to which water would rise in a glass tube whose internal diameter is equal to the distance between the plates.

If the same plates are plunged into mercury, there will be a depression according to an analagous law.

If two plates of glass, AB and AC, inclined to each other, as shown in Fig. 160, their line of junction being vertical, be plunged into any liquid which will moisten them, the liquid will rise between them. It will rise higher near the junction, the surface taking a curved form, such that any section made by a plane through A, will be an equilateral hyperbola. This form of the elevated fluid conforms to the laws above explained.

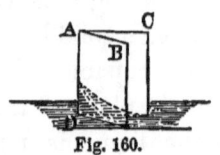

Fig. 160.

If the line of junction of the two plates is horizontal, a small quantity of a liquid between them, which will moisten them, will assume the shape shown at A. If the liquid does not moisten the plates, it will take the form shown at B.

Fig. 161.

Attraction and Repulsion of Floating Bodies.

183. If two small balls of wood, both of which can be moistened by water, or two small balls of wax, which cannot be moistened by water, be placed in a vessel of water, and brought so near each other that the surfaces of capillary elevation or depression interfere, the balls will attract each other and come together. If one ball of wood and one of wax be brought so near that the surfaces of capillary elevation and depression interfere, the bodies will repel each other and separate. If two needles be carefully oiled and laid upon the surface of a vessel of water, they will repel the water from their neighborhood, and float. If, whilst floating, they are brought sufficiently near to each other to permit the surfaces of capillary depression to interfere, the needles will immediately rush together. The reason of the needles floating is, that they repel the water, heaping it up on each side, thus forming a cavity in the surface; the needle is buoyed up by a force equal to the weight of the displaced fluid, and, when this exceeds the weight of the needle, it will float. It is on this principle that certain insects move freely over the surface of a sheet of water; their feet are lubricated with an oily substance which repels the water from around them, producing a hollow around each foot, and giving rise to a buoyant effort greater than the weight of the insect.

The principle of mutual attraction between bodies, both of which repel water, or both of which attract it, accounts for the fact that small floating bodies have a tendency to collect in groups about the borders of the containing vessel. When the material of which the vessel is made, exercises a different capillary action from that of the floating particles, they will aggregate themselves at a distance from the surface of the vessel.

Applications of the Principles of Capillarity.

184. It is in consequence of capillary action that water rises to fill the pores of a sponge, or of a lump of sugar. The same principle, causes the oil to rise in the wick of a lamp, which is but a bundle of fibres very nearly in contact, leaving capillary interstices between them.

The siphon filter differs but little in principle from the wick of a lamp. It consists of a bundle of fibres like a lamp-wick, one end of which dips into a vessel of the liquid to be filtered, whilst the other hangs over the edge of the vessel. The liquid ascends the fibrous mass by the principle of capillary attraction, and continues to advance till it reaches the overhanging end, when, if this is lower than the upper surface of the liquid, the liquid will fall by drops from the end of the wick, the impurities being left behind.

The principle of capillary attraction is used for splitting rocks and raising weights. To employ this principle in cleaving mill-stones, as is done in France, the stone is first dressed to the form of a cylinder of the required diameter for the mill-stone. Grooves are then cut around it where the divisions are to take place, and into these grooves thoroughly dried wedges of willow-wood are driven. On being exposed to the action of moisture, the cells of the wood absorb a large quantity of water, expand, and finally split the rock.

To raise a weight, let a thoroughly dry cord be fastened to the weight, and then stretched to a point above. If, now, the cord be moistened, the fibres will absorb the moisture, expanding laterally, the rope will be diminished in length, and the weight raised.

The principle of capillary attraction is also very extensively employed in metallurgy, in a process of purifying metals, called cupellation.

Endosmose and Exosmose.

185. The names *endosmose* and *exosmose* have been given to two currents flowing in a contrary direction between two liquids, when they are separated by a thin porous partition, either organic or inorganic. The discovery of this phenomena is due to M. DUTROCHET, who called the flowing in, endosmose, and the flowing out, exosmose. The existence of the currents was established by means of an instrument, to which he gave the name *endosmometre*. This instrument consists of a long tube of glass, at one end of which is attached a membranous sack, secured by a tight ligature. If the sack is filled with gum water, a solution of sugar, albumen, or, in fact, with almost any solution denser than water, and then plunged into water, it is observed, after a time, that the fluid rises in the stem, and is depressed in the vessel, showing that water has entered the sack by passing through the pores. By applying suitable tests, it is also found, that a portion cf the liquid in the sack has passed through the pores into the vessel.

Two currents are thus established. If the operation be reversed, and the bladder and tube be filled with pure water, the liquid in the vessel will rise, whilst that in the tube falls. The phenomena of endosmose and exosmose are extremely various, and serve to explain a great variety of interesting facts in animal and vegetable physiology. The cause of the currents is the action of molecular forces exerted between the particles of the bodies employed.

CHAPTER VIII.

MECHANICS OF GASES AND VAPORS.

Gases and Vapors.

186. Gases and vapors are distinguished from other fluids, by their great compressibility, and correspondingly great elasticity. These fluids continually tend to occupy a greater space; this expansion goes on till counteracted by some extraneous force, as that of gravity, or the resistance offered by a containing vessel.

The force of expansion, which is common to all gases and vapors, is called their *tension* or *elastic force*. We shall take for the unit of this force at any point, the pressure which would be exerted upon a square inch of surface, were the pressure the same at every point of the square inch as at the point in question. If we denote this unit, by p, the area pressed, by a, and the entire pressure, by P, we shall have,

$$P = ap \quad \cdots \quad (151.)$$

Most of the principles already demonstrated for liquids hold good for gases and vapors, but there are certain properties arising from elasticity which are peculiar to æriform fluids, some of which it is now proposed to investigate.

Atmospheric Air.

187. The gaseous fluid which envelops our globe, and extends on all sides to a distance of many miles, is called the *atmosphere*. It consists principally of nitrogen and oxygen, together with variable, but small portions of watery vapor and carbonic acid, all in a state of mixture. On an average, it is found by experiment that 1000 parts by volume of

286 MECHANICS.

atmospheric air, taken near the surface of the earth, consists of about,

>788 parts of nitrogen,
>197 parts of oxygen,
>14 parts of watery vapor,
>1 part of carbonic acid.

The atmosphere may, physically speaking, be taken as a type of gases, for it is found by experiment that the laws regulating the density, expansibility, and elasticity, are the same for all gases and vapors, so long as they maintain a purely gaseous form. It is found, however, in the case of vapors, and of those gases which have been reduced to a liquid form, that the law changes just before actual liquefaction.

This change appears to be somewhat analagous to that observed when water passes from the liquid to the solid form. Although water does not actually freeze till reduced to a temperature of 32° Fah., it is found that it reaches its maximum density at about 38°.75, at which temperature the particles seem to commence arranging themselves according to some new laws, preparatory to taking the solid form.

Atmospheric Pressure.

188. If a tube, 35 or 36 inches long, open at one end and closed at the other, be filled with pure mercury, and inverted in a basin of the same, it is observed that the mercury will fall in the tube until the vertical distance from the surface of the mercury in the tube to that in the basin is about 30 inches. This column of mercury is sustained by the pressure of the atmosphere exerted upon the surface of the mercury in the basin, and transmitted through the fluid, according to the general law of *transmission of pressures*. The column of mercury sustained by the elasticity of the atmosphere is called the *barometric column*,

Fig. 162.

because it is generally measured by an instrument called a barometer. In fact, the instrument just described, when provided with a suitable scale for measuring the altitude of the column, is a complete barometer. The height of the barometric column fluctuates somewhat, even at the same place, on account of changes of temperature, and other causes yet to be considered.

Observation has shown, that the average height of the barometric column at the level of the sea, is a trifle less than 30 inches.

The weight of a column of mercury 30 inches in height, having a cross section of one square inch, is nearly 15 pounds. Hence, the unit of atmospheric pressure at the level of the sea, is 15 pounds.

This unit is called an *atmosphere*, and is often employed in estimating the pressure of elastic fluids, particularly in the case of steam. Hence, to say that the pressure of steam in a boiler is two atmospheres, is equivalent to saying, that there is a pressure of 30 pounds upon each square inch of the interior of the boiler. In general, when we say that the tension of a gas or vapor is n atmospheres, we mean that each square inch is pressed by a force of n times 15 pounds.

Mariotte's Law.

189. When a given mass of any gas or vapor is compressed so as to occupy a smaller space, other things being equal, its elastic force is increased; on the contrary, if its volume is increased, its elastic force is diminished.

The law of increase and diminution of elastic force, first discovered by MARIOTTE, and bearing his name, may be enunciated as follows:

The elastic force of a given mass of any gas, whose temperature remains the same, varies inversely as the volume which it occupies.

As long as the mass remains the same, the density must vary inversely as the volume occupied. Hence, from MARIOTTE'S Law, it follows, that,

The elastic force of any gas, whose temperature remains the same, varies as its density, and conversely, *the density varies as the elastic force.*

MARIOTTE's law may be verified in the case of atmospheric air, by the aid of an instrument called MARIOTTE's Tube. This instrument consists of a tube $ABCD$, of uniform bore, bent so that its two branches are parallel to each other. The shorter branch AB, is closed at its upper extremity, whilst the longer one remains open for the reception of mercury. Between the two branches of the tube, and attached to the same frame with it, is a scale of equal parts for measuring distances.

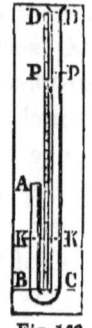

Fig. 168

To use the instrument, place it in a vertical position, and pour mercury into the tube, until it just cuts off the communication between the two branches. The mercury will then stand at the same level BC, in both branches, and the tension of the confined air in AB, will be exactly equal to that of the external atmosphere. If an additional quantity of mercury be poured into the longer branch, the confined air in the shorter branch will be compressed, and the mercury will rise in both branches, but higher in the longer, than in the shorter one. Suppose the mercury to have risen in the shorter branch, to K, and in the longer one, to P. There will be an equilibrium in the mercury lying below the horizontal plane KK; there will also be an equilibrium between the tension of the air in AK, and the forces which give rise to that tension. These forces are the pressure of the external atmosphere transmitted through the mercury, and the weight of a column of mercury whose base is the cross-section of the tube, and whose altitude is PK. If we denote the height of the column of mercury which will be sustained by the pressure of the external atmosphere, by h, the tension of the air in AK, will be measured by the weight of a column of mercury, whose base is the cross-section of the tube, and whose height is $h + PK$. Since the weight is proportional to the height, the tension of the confined air will be proportional to $h + PK$.

Now, whatever may be the value of PK, it is found that,

$$AK = \frac{\overline{AB} \cdot h}{h + PK}.$$

If $PK = h$, we shall have, $AK = \frac{1}{2}AB$; if $PK = 2h$, we shall have, $AK = \frac{1}{3}AB$; in general, if $PK = nh$, n being any positive number, either entire or fractional, we shall have, $AK = \dfrac{AB}{n+1}$. Mariotte's Law was verified in this manner by Dulong and Arago for all values of n, up to $n = 27$. The law may also be verified when the pressure is less than an atmosphere, by means of the following apparatus.

AK represents a straight tube of uniform bore, closed at its upper and open at its lower extremity: CD is a long cistern of mercury. The tube AK is either graduated into equal parts, commencing at A, or it has attached to it a scale of brass or ivory.

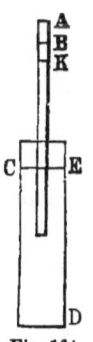

Fig. 164.

To use the instrument, pour mercury into the tube till it is nearly full; place the finger over the open end, and invert it in the cistern of mercury, and depress it till the mercury stands at the same level without, as within the tube, and suppose the surface of the mercury in this case to cut the tube at B. Then will the tension of the confined air in AB, be equal to that of the external atmosphere. If now the tube be raised vertically, the air in AB will expand, its tension will diminish and the mercury will fall in the tube, to maintain the equlibrium. Suppose the level of the mercury in the tube to have reached the point K. In this position of the instrument the tension of the air in AK, added to the weight of the column of mercury, KE will be equal to the tension of the external air.

Now, it is found, whatever may be the value of KE, that

$$AK = \frac{\overline{AB} \cdot h}{h - EK}.$$

If $EK = \frac{1}{2}h$, we have, $AK = 2AB$; if $EK = \frac{2}{3}h$, we have, $AK = 3AB$; in general, if $EK = \frac{n}{n+1}h$, we have, $AK = \frac{AB}{n+1}$.

Mariotte's law has been verified in this manner, for all values of n, up to $n = 111$.

It is a law of Physics that, when a gas is suddenly compressed, heat is evolved, and when a gas is suddenly expanded, heat is absorbed; hence, in making the experiment, care must be taken to have the temperature kept uniform.

Gay Lussac's Law.

190. If, whilst the volume of any gas or vapor remains the same, its temperature be increased, its tension is increased also. If the pressure remain the same, the volume of the gas will increase as the temperature is raised. The law of increase and diminution, as deduced by Gay Lussac, whose name it bears, may be enunciated as follows:

In a given mass of any gas, or vapor, if the volume remains the same, the tension varies as the temperature; if the tension remains the same, the volume varies as the temperature.

According to Regnault, if a given mass of atmospheric air be heated from 32° Fahrenheit to 212°, the tension, or pressure remaining constant, its volume will be increased by the .3665th part of the volume at 32°. Hence, the increase of volume for each degree of temperature is the .00204th part of the volume at 32°. If we denote the volume at 32° by v, and the volume at the temperature t', by v', we shall therefore have,

$$v' = v[1 + .00204(t' - 32)] \quad . \quad . \quad (152.)$$

Solving with reference to v, we have,

$$v = \frac{v'}{1 + .00204(t' - 32)} \quad . \quad . \quad . \quad (153.)$$

Formula (153) enables us to compute the volume of any

mass of air at 32°, knowing its volume at the temperature t', the pressure remaining constant.

To find the volume at the temperature t'', we have simply to substitute t'' for t' in (152.) Denoting this volume by v'', we have,

$$v'' = v[1 + .00204(t'' - 32)].$$

Substituting for v its value from (153), we get,

$$v'' = v' \frac{1 + .00204(t'' - 32)}{1 + .00204(t' - 32)} \quad \cdot \quad \cdot \quad (154.)$$

This formula enables us to compute the volume of any mass of air, at a temperature t'', when we know its volume at the temperature t'; and, since the density varies inversely as the volume, we may also, by means of the same formula, find the density of any mass of air, at the temperature t'', when we have given its density at the temperature t'.

Manometers.

191. A MANOMETER is an instrument used for measuring the tension of gases and vapors, and particularly of steam. Two principle varieties of manometers are used for measuring the tension of steam, the *open manometer*, and the *closed manometer*.

The open Manometer.

192. The open manometer consists, essentially, of an open glass tube AB, terminating below, nearly at the bottom of a cistern EF. The cistern is of wrought iron, steam tight, and filled with mercury. Its dimensions are such, that the upper surface of the mercury will not be materially lowered, when a portion of the mercury is forced up the tube. ED is a tube, by means of which, steam may be admitted from the boiler to the surface of the mercury in the cistern. This tube is sometimes filled with

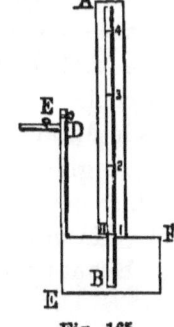

Fig. 165.

water, through which the pressure of the steam is transmitted to the mercury.

To graduate the instrument. All communication with the boiler is cut off, by closing the stop-cock E, and communication with the external air is made by opening the stopcock D. The point of the tube AB, to which the mercury rises, is noted, and a distance is laid off, upwards, from this point, equal to what the barometric column wants of 30 inches, and the point H thus determined, is marked 1. This point will be very near the surface of the mercury in the cistern. From the point H, distances of 30, 60, 90, &c., inches are laid off upwards, and the corresponding points numbered 2, 3, 4, &c. These divisions correspond to atmospheres, and may be subdivided into tenths and hundredths.

To use the instrument, the stop-cock D is closed, and a communication made with the boiler, by opening the stop-cock E. The height to which the mercury rises in the tube, will indicate the tension of the steam in the boiler, which may be read from the scale in terms of atmospheres and decimals of an atmosphere. If the pressure in pounds is wished, it may at once be found, by multiplying the reading of the instrument by 15.

The principal objection to this kind of manometer, is its want of portability, and the great length of tube required, when high tensions are to be measured.

The closed Manometer.

295. The general construction of the closed manometer is the same as that of the open manometer, with the exception that the tube AB is closed at the top. The air which is confined in the tube, is then compressed in the same way as in MARIOTTE's tube.

To graduate this instrument. We determine the division H, as before. The remaining divisions are found by applying MARIOTTE's law.

Denote the distance in inches, from H to the top of the

tube, by l; the pressure on the mercury, expressed in atmospheres, by n, and the distance in inches, from H to the upper surface of the mercury in the tube, by x.

The tension of the air in the tube will be equal to that on the mercury in the cistern, diminished by the weight of a column of mercury, whose altitude is x. Hence, in atmospheres, it is

$$n - \frac{x}{30}.$$

The bore of the tube being uniform, the volume occupied by the compressed air will be proportional to its height. When the pressure is 1 atmosphere, the height is l; when the pressure is $n - \frac{x}{30}$ atmospheres, the height is $l - x$. Hence, from MARIOTTE's law,

$$1 : n - \frac{x}{30} :: l - x : l.$$

Whence, by reduction,

$$x^2 - (30n + l)x = -30l(n - 1).$$

Solving, with respect to x, we have,

$$x = \frac{30n + l}{2} \pm \sqrt{-30l(n-1) + \left(\frac{30n+l}{2}\right)^2}.$$

The upper sign of the radical is not used, as it would give a value for x, greater than l. Taking the lower sign, and, as a particular case, assuming $l = 30$ in., we have,

$$x = 15n + 15 - \sqrt{-900(n-1) + (15n+15)^2}.$$

Making $n = 2, 3, 4,$ &c., in succession, we find for x, the corresponding values, 11.46 in., 17.58 in., 20.92 in., &c. These distances being set off from H, upwards, and marked 2, 3, 4, &c., indicate atmospheres. The intermediate spaces are subdivided by means of the same formula.

The use of this instrument is the same as that of the manometer last described.

In making the graduation, we have supposed the temperature to remain the same. If, however, it does not remain the same, the reading of the instrument must be corrected by means of a table computed for the purpose.

The instruments already described, can only be used for measuring tensions greater than one atmosphere.

The Siphon Guage.

194. The SIPHON GUAGE is an instrument employed to measure tensions of gases and vapors, when they are less than an atmosphere. It consists of a tube ABC, bent so that its two branches are parallel. The branch BC is closed at the top, and filled with mercury, which is retained by the pressure of the atmosphere, whilst the branch AB is open at the top. If, now, the air be rarified in any manner, or if the mouth A of the tube, be exposed to the action of any gas whose tension is sufficiently small, the mercury will no longer be supported in the branch BC, but will fall in that and rise in the other. The distance between the surfaces of the mercury in the two branches, as given by a scale placed between them, will indicate the tension of the gas. If this distance is expressed in inches, the tension can be found, in atmospheres, by dividing by 30, or, in pounds, by dividing by 2.

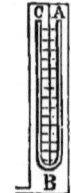

Fig. 166.

The Diving-Bell.

195. The DIVING-BELL is a bell-shaped vessel, open at the bottom, used for descending below the surface of the water. The bell is placed so that its mouth shall continue horizontal, and is let down by means of a rope AB, and the whole apparatus is sunk by weights properly adjusted. The air contained in the bell before immersion, will be compressed by the weight of the

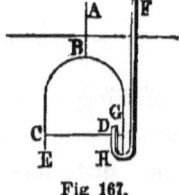

Fig 167.

water, but its increased elasticity will prevent the water from rising to the top of the bell, which is provided with seats for the accommodation of those wishing to descend. The air within is constantly contaminated by breathing, and is continually replaced by fresh air, pumped in through a tube *FG*. Were there no additional air introduced, the volume of the compressed air, at any depth, might be computed by MARIOTTE's law. The unit of the compressing force, in this case, is the weight of a column of water whose cross-section is a square inch, and whose height is the distance from *DC*, to the surface of the water.

The Barometer.

196. The BAROMETER is an instrument for measuring the pressure of the atmosphere. As already explained, it consists of a glass tube, hermetically sealed at one extremity, which is filled with mercury, and inverted in a basin of that fluid. The pressure of the air is indicated by the height of the column of mercury which it supports.

A great variety of forms of the mercurial barometer have been devised, all involving the same mechanical principle. The two most important of these are the *siphon* and the *cistern barometer*.

The Siphon Barometer.

197. The siphon barometer consists essentially of a tube *CDE*, bent so that its two branches, *CD* and *DE*, shall be parallel to each other. A scale of equal parts is placed between them, and attached to the same frame with the tube. The longer branch *CD*, is about 32 or 33 inches in length, hermetically sealed at the top, and filled with mercury; the shorter one is open to the action of the air. When the instrument is placed vertically, the mercury sinks in the longer branch and rises in the shorter one. The distance between the surface of the mercury in the two branches, as measured by the scale of equal parts, indicates the pressure of the atmosphere at the particular time and place.

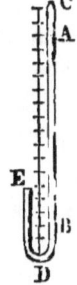

Fig. 168.

The Cistern Barometer.

198. The cistern barometer consists of a glass tube, filled and inverted in a cistern of mercury, as already explained. The tube is surrounded by a frame of metal, firmly attached to the cistern. Two opposite longitudinal openings, near the upper part of the frame, permit the upper surface of the mercury to be seen. A slide, moved up and down by means of a rack and pinion, may be brought exactly to the upper level of the mercury. The height of the column is then read from a scale, so adjusted as to have its 0 at the surface of the mercury in the cistern. The scale is graduated to inches and tenths, and the smaller divisions are read by means of a vernier.

The figure shows the arrangement of parts in a complete cistern barometer. *KK* represents the frame of the barometer; *HH* that of the cistern, open at the upper part, that the level of the mercury in the cistern may be seen through the glass; *L*, an attached thermometer, to show the temperature of the mercury in the tube; *N*, a part of the sliding ring bearing the vernier, and moved up and down by the milled-headed screw *M*.

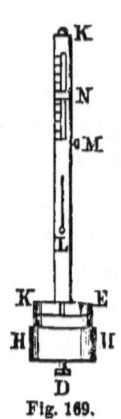
Fig. 169.

The particular arrangement of the cistern is shown on an enlarged scale in Fig. 170. *A* represents the barometer tube, terminating in a small opening, to prevent too sudden shocks when the instrument is moved from place to place; *H* represents the frame of the cistern; *B*, the upper portion of the cistern, made of glass, that the surface of the mercury may be seen; *E*, a conical piece of ivory, projecting from the upper surface of the cistern: when the surface of the mercury just touches the point of the ivory, it is at the 0 of the scale; *CC* represents the lower part of the cistern, and is made of leather, or some other

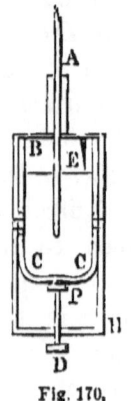
Fig. 170.

flexible substance, and firmly attached to the glass part. D is a screw, working through the bottom of the frame, and against the bottom of the bag CC, through the medium of a plate P. The screw D, serves to bring the surface of the mercury to the point of the ivory piece E, and also to force the mercury up to the top of the tube, when it is desired to transport the barometer from place to place.

To use this barometer, it should be suspended vertically, and the level of the mercury in the cistern brought to the point of the ivory piece E, by means of the screw D; a smart rap with a key upon the frame will detach the mercury from the glass to which it sometimes tends to adhere. The sliding ring N, is next run up or down by means of the screw M, till its lower edge appears tangent to the upper surface of the mercury in the tube, and the altitude is read from the scale. The height of the attached thermometer should also be noted.

The requirements of a good barometer are, sufficient width of tube, perfect purity of the mercury, and a scale with a vernier accurately graduated and adjusted.

The bore of the tube should be as large as practicable, to diminish the effect of capillary action. On account of the mutual repulsion between the particles of the glass and mercury, the mercury is depressed in the tube, and this depression increases as the diameter of the tube diminishes.

In all cases, this depression should be allowed for, and corrected by means of a table computed for the purpose.

To secure purity of the mercury, it should be carefully distilled, and after the tube is filled, it should be boiled over a spirit-lamp, to drive off any bubbles of air that might adhere to the walls of the tube.

Uses of the Barometer.

199. The primary object of the barometer is, to measure the pressure of the atmosphere at any time or place. It is used by mariners and others, as a weather-glass. It is also extensively employed for determining the heights of points on the earth's surface, above the level of the ocean.

The principle on which it is employed for the latter purpose is, that the pressure of the atmosphere at any place depends upon the weight of a column of air reaching from the place to the upper limit of the atmosphere. As we ascend above the level of the ocean, the weight of the column diminishes; consequently, the pressure becomes less, a fact which is shown by the mercury falling in the tube. We shall investigate a formula for determining the difference of level between any two points.

Difference of Level.

200. Let aB represent a portion of a vertical prism of air, whose cross-section is one square inch. Denote the pressure on the lower base B, by p, and on the upper base aa', by p'; denote the density of the air at B, by d, and at aa', by d', and suppose the temperature throughout the column to be 32° Fah.

Pass a horizontal plane bb', infinitely near to aa', and denote the weight of the elementary volume of air ab, by w. Conceive the entire column to be divided by horizontal planes into elementary prisms, such that the weights of each shall be equal to w, and denote their heights, beginning at a, by s, s', s'', &c.

Fig. 171.

From MARIOTTE'S law, we shall have,

$$\frac{p'}{p} = \frac{d'}{d}.$$

The air throughout each elementary prism may be regarded as homogeneous; hence, the density of the air in ab is equal to its weight, divided by its volume into gravity (Art. 12). But its volume is equal to $1 \times 1 \times s = s$ hence,

$$d' = \frac{w}{gs}.$$

Substituting this in the preceding equation, we have,

$$\frac{p'}{p} = \frac{w}{gsd};$$

whence,

$$s = \frac{p}{dg} \times \frac{w}{p'} \quad \cdots \quad (155.)$$

From DAVIES' Bourdon, page 297, we have, by substituting for y the fraction $\frac{w}{p'}$, the equation,

$$l\left(1 + \frac{w}{p'}\right) = \frac{w}{p'} - \frac{w^2}{2p'^2} + \frac{w^3}{3p'^3} - \&c.$$

But $\frac{w}{p'}$ being infinitely small, all the terms in the second member, after the first, may be neglected, giving,

$$\frac{w}{p'} = l\left(1 + \frac{w}{p'}\right); \quad \text{or,} \quad \frac{w}{p'} = l\left(\frac{p'+w}{p'}\right);$$

or finally,

$$\frac{w}{p'} = l(p' + w) - lp',$$

in which l denotes the Napierian logarithm.

In this equation, p' denotes the pressure on the prism ab; hence, $p' + w$ denotes the pressure on the next prism below, that is, on the prism bc.

If we substitute this value of $\frac{w}{p'}$ in Equation (155), we shall have, for the height of the prism ab,

$$s = \frac{p}{dg}[l(p' + w) - lp'].$$

Substituting in succession for p', the values $p' + w$, $p' + 2w$, $p' + 3w$, &c., we shall find the heights of the elementary prisms bc, cd, &c. We shall therefore have,

$$s = \frac{p}{dg}[l(p'+w) - lp'],$$

$$s' = \frac{p}{dg}[l(p'+2w) - l(p'+w)],$$

$$s'' = \frac{p}{dg}[l(p'+3w) - l(p'+2w)],$$

. . . .

$$s^{n'} = \frac{p}{dg}[l(p'+nw) - l(p'+(n-1)w)].$$

If n denote the number of elementary prisms in AB, the sum of the first members will be equal to AB. Adding the equations member to member, and denoting the sum of the first members by z, we have,

$$z = \frac{p}{dg}[l(p'+nw) - lp'] = \frac{p}{dg} \times l\left(\frac{p'+nw}{p'}\right).$$

Because nw denotes the weight of the column of air AB, we shall have, $p' + nw = p$, hence,

$$z = \frac{p}{dg} l\frac{p}{p'}. \quad \cdots \quad (156.)$$

Denoting the modulus of the common system of logarithms by M, and designating common logarithms by the symbol log, we shall have,

$$Mz = \frac{p}{dg} \log \frac{p}{p'}, \quad \text{or} \quad z = \frac{p}{Mdg} \log \frac{p}{p'}.$$

Now, the pressures p and p' are measured by the heights of the columns of mercury which they will support; denoting these heights by H and H', we have,

$$\frac{p}{p'} = \frac{H}{H'}.$$

whence, by substitution,

$$z = \frac{p}{Mdg} \log \frac{H}{H'}. \quad \cdots \quad (157.)$$

We have supposed the temperature, both of the air and mercury, to be 32°. In order to make the preceding formula general, let T represent the temperature of the mercury at B, T'', its temperature at a, and denote the corresponding heights of the barometric column by h and h'; also, let t denote the temperature of the air at B, and t' its temperature at a.

The quantity $\frac{p}{d}$ is the ratio of the density of the air at B, to the corresponding pressure, the temperature being 32°. According to MARIOTTE's law, this ratio remains constant, whatever may be the altitude of B above the level of the ocean.

If we denote the latitude of the place by l, we have, (Art. 124),

$$g = g'(1 - 0.002695 \cos 2l)$$

It has been shown, by experiment, that, when a column of mercury is heated, it increases in length at the rate of $\frac{1}{9990}$ths of its length at 32°, for each degree that the temperature is elevated. Hence,

$$h = H\left(1 + \frac{T - 32}{9990}\right) = H \frac{9990 + T - 32}{9990};$$

$$h' = H'\left(1 + \frac{T' - 32}{9990}\right) = H' \frac{9990 + T'' - 32}{9990}.$$

Dividing the second equation by the first, member by member,

$$\frac{h}{h'} = \frac{H}{H'} \cdot \frac{9990 + T - 32}{9990 + T'' - 32}.$$

Dividing both terms of the fractional coefficient of $\dfrac{H}{H'}$ by the denominator, and neglecting the quantity $T-32$, in comparison with 9990, we have,

$$\frac{h}{h'} = \frac{H}{H'}\left(1 + \frac{T-T''}{9990}\right) = \frac{H}{H'}(1+.0001)(T-T').$$

Whence, by reduction,

$$\frac{H}{H'} = \frac{h}{h'}\cdot\frac{1}{1+.0001(T-T'')}.$$

The quantity z denotes, not only the height, but also the volume of the column of air aB, at 32°. When the temperature is changed from 32°, the pressures remaining the same, this volume will vary, according to the law of GAY LUSSAC.

If we suppose the temperature of the entire column to be a mean between the temperatures at B and a, which we may do without sensible error, the height of the column will become, Equation (153),

$$z\left[1 + .00204\left(\frac{t+t'}{2} - 32\right)\right] = z[1+.00102(t+t'-64)]$$

Hence, to adapt Equation (157) to the conditions proposed, we must multiply the value of z by the factor,

$$1 + .00102(t + t' - 64).$$

Substituting in Equation (157), for $\dfrac{H}{H'}$ and g, the values shown above, and multiplying the resulting value of z, by the factor $1 + .00102(t + t' - 64)$, we have,

$$z = \frac{p}{Md}\cdot\frac{1 + .00102(t+t'-64)}{1 - 0.002695\cos 2l}\log\frac{h}{h'[1+.0001(T-T'')]}$$
(158.)

The factor $\dfrac{p}{Md}$ is constant, and may be determined as follows: select two points, one of which is considerably higher than the other, and determine, by trigonometrical measurement, their difference of level. At the lower point, take the reading of the barometer, of its attached thermometer, and of a detached thermometer exposed to the air. Make similar observations at the upper station. These observations, together with the latitude of the place, will give all the quantities entering Equation (158), except the factor in question. Hence, this factor may be deduced. It is found to be 60345.51 ft. Hence, we have, finally, the barometric formula,

$$z = 60345.51 \text{ ft.} \times \frac{1 + .00102\,(t+t' - 64)}{1 - 0.002695\cos 2l} \log \frac{h}{h'[1 + .0001(T - T')]} \quad (159.)$$

To use this formula for determining the difference of level between two stations, observe, simultaneously, if possible, the heights of the barometer and of the attached and detached thermometers, at the two stations. Substitute these results for the corresponding quantities in the formula; also substitute for l the latitude of the place, and the resulting value of z, will be the difference of level required.

If the observations cannot be made simultaneously at the two stations, make a set of observations at the lower station; after a certain interval, make a set at the upper station; then, after an equal interval, make another set at the lower station. Take a mean of the results of observation at the lower station, as a single set, and proceed as before.

For the more convenient application of the formula for the difference of level between two points, tables have been computed, by means of which the arithmetical operations are much facilitated.

Work due to the Expansion of a Gas or Vapor.

201. Let the gas or vapor be confined in a cylinder closed at its lower end, and having a piston working air-tight. When the gas occupies a portion of the cylinder whose height is h, denote the pressure on each square inch of the piston by p; when the gas expands, so that the altitude of the column becomes x, denote the pressure on a square inch by y.

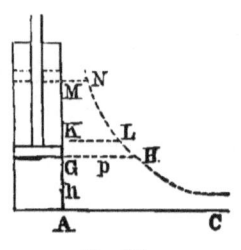

Fig. 172.

Since the volumes of the gas, under these suppositions, are proportional to their altitudes, we shall have, from Mariotte's laws,

$$p : y :: x : h;$$

whence

$$xy = ph$$

If we suppose p and h to be constant, and x and y to vary, the above equation will be that of an equilateral hyperbola referred to its asymptotes.

Draw AC perpendicular to AM, and on these lines, as asymptotes, construct the curve NLH, from the equation, $xy = ph$. Make $AG = h$, and draw GH parallel to AC; it will represent the pressure p. Make $AM = x$, and draw MN parallel to AC; it will represent the pressure y. In like manner, the pressure at any elevation of the piston may be constructed.

Let KL be drawn infinitely near to GH, and parallel with it. The elementary area $GKLH$ will not differ sensibly from a rectangle whose base is p, and altitude is GK. Hence, its area may be taken as the measure of the work whilst the piston is rising through the infinitely small space GK. In like manner, the area of any infinitely small element, bounded by lines parallel to AC, may be taken to represent the work whilst the piston is rising through the height of the element. If we take the sum of all the elements between the ordinates GH and MN, this sum, or the area $GMNH$, will represent the total quantity of work of the force of expansion whilst the piston is rising from G to M. But the area included between an equilateral hyperbola and one of its asymptotes, and limited by lines parallel to the other asymptote, is equal to the product of the co-ordinates of any point, multiplied by the Naperian logarithm of the quotient obtained by dividing one of the limiting ordinates by the other; or, in this particular case, it is equal to $ph \times l\left(\frac{p}{y}\right)$. Hence, if we designate the quantity of work performed by the expansive force whilst the piston is moving over GM, by q, we shall have,

$$q = ph \times l\left(\frac{p}{y}\right).$$

This is the quantity of work exerted upon each square inch of the piston; if we denote the area of the piston, by A, and the total quantity of work, by Q, we shall have,

$$Q = Aph \times l\left(\frac{p}{y}\right) = Aph \times l\left(\frac{x}{h}\right) \quad . \quad (160.)$$

If we denote by c the number of cubic feet of gas, when the pressure is p, and suppose it to expand till the pressure is y, we shall have, $Ah = c$; or, if A be expressed in square feet, we shall have, $c = \dfrac{Ah}{144}$. Hence, by substitution,

$$Q = 144cp \times l\left(\frac{p}{y}\right).$$

Finally, if we suppose the pressure at the highest point to be p', we shall have,

$$Q = 144cp \times l\left(\frac{p}{p'}\right),$$

an equation which gives the quantity of work of c cubic feet of gas, whilst expanding from a pressure p, to a pressure p'.

Efflux of a Gas or Vapor.

202. Suppose the gas to escape from a small orifice, and denote its velocity by v. Denote the weight of a cubic foot of the gas, by w, and the number of cubic feet discharged in one second, by c, then will the mass escaping in one second, be equal to $\dfrac{cw}{g}$, and its living force will be equal to $\dfrac{cw}{g} v^2$. But, from Art. 148, the living force is double the accumulated quantity of work. If, therefore, we denote the accumulated work by Q, we shall have,

$$Q = \frac{cw}{2g} v^2.$$

But the accumulated work is due to the expansion of the gas, and if we denote the pressure within the orifice, by p, and without, by p', we shall have, from Art. 201,

$$Q = 144 cp \times l\!\left(\frac{p}{p'}\right).$$

Equating the second members, we have,

$$\frac{cw}{2g} v^2 = 144 cp \times l\!\left(\frac{p}{p'}\right);$$

Whence,

$$v = 12 \sqrt{\frac{2gp}{w} \times l\!\left(\frac{p}{p'}\right)}.$$

Substituting for g, its value, $32\tfrac{1}{6}$ ft., we have, after reduction,

$$v = 96 \sqrt{\frac{p}{w} \times l\!\left(\frac{p}{p'}\right)} \quad \ldots \quad (161.)$$

When the difference between p and p' is small, the preceding formula can be simplified.

Since $\dfrac{p}{p'} = 1 + \dfrac{p-p'}{p'}$, we have, from the logarithmic series,

$$l\left(\frac{p}{p'}\right) = l\left(1 + \frac{p-p'}{p'}\right) = \frac{p-p'}{p'} - \tfrac{1}{2}\left(\frac{p-p'}{p'}\right)^2 + \&c.$$

When $p - p'$ is very small, the second, and all succeeding terms of the development, may be neglected, in comparison with the first term. Hence,

$$l\left(\frac{p}{p'}\right) = \frac{p-p'}{p'}.$$

Substituting, in the formula above deduced, we have,

$$v = 96\sqrt{\frac{p}{w} \times \frac{p-p'}{p'}};$$

or, since $\dfrac{p}{p'}$ is, under the supposition just made, equal to 1, we have, finally,

$$v = 96\sqrt{\frac{p-p'}{w}} \quad \ldots \quad (162.)$$

Coefficient of Efflux.

203. When air issues from an orifice, the section of the current undergoes a change of form, analagous to the contraction of the vein in liquids, and for similar reasons. If we denote the coefficient of efflux, by k, the area of the orifice, by A, and the quantity of air delivered in n seconds, by Q, we shall have, from Equation (161),

$$Q = 96knA\sqrt{\frac{p}{w}\, l\left(\frac{p}{p'}\right)}.$$

According to Koch, the value of k is equal to .58, when the orifice is in a thin plate; equal to .74, when the air issues through a tube 6 times as long as it is wide; and equal to .85, when it issues through a conical nozzle 5 times as long as the diameter of the orifice, and whose sides have a convergence of 6° to the axis.

The preceding principles are applicable to the distribution of gas, to the construction of blowers, and, in general, to a great variety of pneumatic machines.

Steam.

204. If water be exposed to the atmosphere, at ordinary temperatures, a portion is converted into vapor, which mixes with the atmosphere, constituting one of the permanent elements of the aerial ocean. The tension of watery vapor thus formed, is very slight, and the atmosphere soon ceases to absorb any more. If the temperature of the water be raised, an additional amount of vapor is evolved, and of greater tension. When the temperature is raised to that point at which the tension of the vapor is equal to that of the atmosphere, ebullition commences, and the vaporization goes on with great rapidity. If heat be added beyond the point of ebullition, neither the water nor the vapor will increase in temperature till all of the water is converted into steam. When the barometer stands at 30 inches, the boiling point of pure water is 212° Fah. We shall suppose, in what follows, that the barometer stands at 30 inches. After the temperature of the water is raised to 212°, the additional heat that is added becomes latent in the vapor evolved.

If heat be applied uniformly, it is found by experiment that it takes $5\frac{1}{2}$ times as much to convert all of the water into steam as it requires to raise it from 32° to 212°. Hence, the entire amount of heat which becomes latent is $5\frac{1}{2} \times (212° - 32°) = 990°$. That the heat applied becomes latent, may be shown experimentally as follows:

Let a cubic inch of water be converted into steam at

212°, and kept in a close vessel. Now, if 5¼ cubic inches of water at 32° be injected into the vessel, the steam will all be converted into water, and the 6¼ cubic inches of water will be found to have a temperature of 212°. The heat that was latent becomes sensible again.

When water is converted into steam under any other pressure than that of the atmosphere, or 15 pounds to the square inch, it is found that, although the boiling point will be changed, the entire amount of heat required for converting the water into steam will remain unchanged.

If the evaporation takes place under such a pressure, that the boiling point is but 150°, the amount of heat which becomes latent is 1052°, so that the latent heat of the steam, plus its sensible heat, is 1202°. If the pressure under which vaporization takes place is such as to raise the boiling point to 500°, the amount of heat which becomes latent is 702°, the sum 702° + 500° being equal to 1202°, as before. Hence, we conclude that *the same amount of fuel is required to convert a given amount of water into steam, no matter what may be the pressure under which the evaporation takes place.*

When water is converted into steam under a pressure of one atmosphere, each cubic inch is expanded into about 1700 cubic inches of steam, of the temperature of 212°; or, since a cubic foot contains 1728 cubic inches, we may say, in round numbers, that *a cubic inch of water is converted into a cubic foot of steam.*

If water is converted into steam under a greater or less pressure than one atmosphere, the density will be increased or diminished, and, consequently, the volume will be diminished or increased. The temperature being also increased or diminished, the increase of density or decrease of volume will not be exactly proportional to the increase of pressure; but, for purposes of approximation, we may consider the densities as directly, and the volumes as inversely proportional to the pressures under which the steam is generated. Under this hypothesis, if a cubic inch of water be evaporated under a pressure of a half atmosphere, it will afford two cubic feet of steam; if generated under a pressure of two atmospheres, it will only afford a half cubic foot of steam.

Work of Steam.

205. When water is converted into steam, a certain amount of work is generated, and, from what has been shown, this amount of work is very nearly the same, whatever may be the temperature at which the water is evaporated.

Suppose a cylinder, whose cross-section is one square inch, to contain a cubic inch of water, above which is an air-tight piston, that may be loaded with weights at pleasure. In the first place, if the piston is pressed down by a weight of 15 pounds, and the inch of water converted into steam, the weight will be raised to the height of 1728 inches, or 144 feet. Hence, the quantity of work is 144 × 15, or, 2160 units. Again, if the piston be loaded with a weight of 30 pounds, the conversion of water into steam will give but 864 cubic inches, and the weight will be raised through 72 feet. In this case, the quantity of work will be 72 × 30, or 2160 units, as before. We conclude, therefore, that the quantity of work is the same, or nearly so, whatever may be the pressure under which the steam is generated. We also conclude, that the quantity of work is nearly proportional to the fuel consumed.

Besides the quantity of work developed by simply converting an amount of water into steam, a further quantity of work is developed by allowing the steam to expand after entering the cylinder. This principle is made use of in steam engines working expansively.

To find the quantity of work developed by steam acting expansively. Let AB represent a cylinder, closed at A, and having an air-tight piston D. Suppose the steam to enter at the bottom of the cylinder, and to push the piston upward to C, and then suppose the opening at which the steam enters, to be closed. If the piston is not too heavily loaded, the steam will continue to expand, and the piston

Fig. 178.

will be raised to some position, B. The expansive force of the steam will obey MARIOTTE's law, and the quantity of work due to expansion will be given by Equation (160).

Denote the area of the piston in square inches, by A; the pressure of the steam on each square inch, up to the moment when the communication is cut off, by p; the distance AC, through which the piston moves before the steam is cut off, by h; and the distance AD, by nh.

If we denote the pressure on each square inch, when the piston arrives at B, by p', we shall have, by MARIOTTE's law,

$$p : p' :: nh : h, \quad \therefore \; p' = \frac{p}{n},$$

an expression which gives the limiting value of the load of the piston.

The quantity of work due to expansion being denoted by q, we shall have, from Equation (160),

$$q = Aph \times l\left(\frac{nh}{h}\right) = Aphl(n).$$

If we denote the quantity of work of the steam, whilst the piston is rising to C, by q'', we shall have,

$$q'' = Aph.$$

Denoting the total quantity of work during the entire stroke of the piston, by Q, we shall have,

$$Q = Aph\left[1 + l(n)\right] \quad . \quad . \quad . \quad (163.)$$

Experimental Formulas.

206. Numerous experiments have been made for the purpose of determining the relation existing between the elasticity and temperature of steam in contact with the water by which it is produced, and many formulas, based upon these experiments, have been given, two of which are subjoined:

The formula of Dulong and Arago is,

$$p = (1 + .007153t)^3,$$

in which p represents the tension in atmospheres, and t the excess of the temperature above 100° Centigrade.

Tredgold's formula is,

$$t = 0.85\sqrt{p} - 75,$$

in which t is the temperature, in degrees of the Centigrade thermometer, and p the pressure, expressed in centimeters of the mercurial column.

CHAPTER IX.

HYDRAULIC AND PNEUMATIC MACHINES.

Definitions.

207. HYDRAULIC MACHINES are those used in raising and distributing water, such as *pumps, siphons, hydraulic rams*, &c. The name is also applied to those machines in which water power is the motor, or in which water is employed to transmit pressures, such as *water-wheels, hydraulic presses*, &c.

PNEUMATIC MACHINES are those employed to rarefy and condense air, or to impart motion to the air, such as *air-pumps, ventilating-blowers*, &c. The name is also applied to those machines in which currents of air furnish the motive power, such as windmills, &c.

Water Pumps.

208. A *water pump* is a machine for raising water from a lower to a higher level, generally by the aid of atmospheric pressure. Three separate principles are employed in the working of pumps: the *sucking*, the *lifting*, and the *forcing* principle. Pumps are frequently named according as one or more of these principles are employed.

Sucking and Lifting Pump.

209. This pump consists of a cylindrical barrel *A*, at the lower extremity of which is attached a sucking-pipe *B*, leading to a reservoir. An air-tight piston *C* is worked up and down in the barrel by means of a lever *E*, attached to a piston-rod *D*. *P* represents a valve opening upwards, which, when the

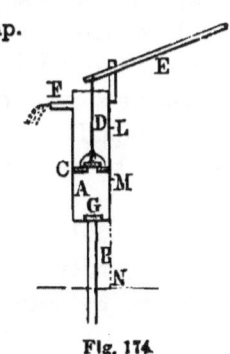

Fig. 174.

pump is at rest, closes by its own weight. This valve is called, from its position, the *piston-valve*. A second valve G, also opening upwards, is placed at the junction of the pipe with the barrel. This is called the *sleeping-valve*. The space LM, through which the piston can be moved up and down by the lever, is called *the play of the piston*.

To explain the action of the pump, suppose the piston to be at the lowest limit of the play, and everything in a state of equilibrium. If the extremity of the lever E be depressed, and the piston consequently be raised, the air in the lower part of the barrel will be rarefied, and that in the pipe B will, by virtue of its greater tension, open the valve, and a portion of it will escape into the barrel. The air in the pipe, thus rarefied, will exert a less pressure upon the water in the reservoir than that of the external air, and, consequently, the water will rise in the pipe, until the tension of the internal air, plus the weight of the column of water raised, is equal to the tension of the external air; the valve G will then close by its own weight.

If the piston be again depressed to the lowest limit, by means of the lever E, the air in the lower part of the barrel will be compressed, its tension will become greater than that of the external air, the valve F will be forced open, and a portion of the air will escape. If the piston be raised once more, the water will, for the same reason as before, rise still higher in the pipe, and after a few double strokes of the piston, the air will be completely exhausted from beneath the piston, the water will pass through the piston valve, and finally escape at the spout P.

The water is raised to the piston by the pressure of the air on the surface of the water in the reservoir; hence, the piston should not be placed at a greater distance above the level of the water in the reservoir, than the height to which the pressure of the air will sustain a column of water. In fact, it should be placed a little lower than this limit. The specific gravity of mercury being about 13.5, the height of a column of water which will exactly counterbalance the pressure of the atmosphere, will be found by multiplying the height of the barometric column by $13\frac{1}{2}$.

At the level of the sea the average height of the barometric column is $2\frac{1}{2}$ feet; hence, the theoretical height to which water can be raised by the principle of suction alone, is a little less than 34 feet.

The water having passed through the piston valve, it may be raised to any height by the lifting principle, the only limitation being the strength of the pump and want of power.

There are certain relations which must exist between the play of the piston and its height above the water in the reservoir, in order that the water may be raised to the piston; for, if the play is too small, it will happen after a few strokes of the piston, that the air between the piston and the surface of the water will not be sufficiently compressed to open the piston valve; when this state of affairs takes place, the water will cease to rise.

To investigate the relation that must exist between the play and the height of the piston above the water.

Denote the play of the piston, by p, the distance from the upper surface of the water in the reservoir to the highest position of the piston, by a, and the height at which the water ceases to rise in the pump, by x. The distance from the surface of the water in the pump to the highest position of the piston will then be equal to $a - x$, and the distance to the lowest position of the piston, will be $a - p - x$. Denote the height at which the atmospheric pressure will sustain a column of water in vacuum, by h, and the weight of a column of water, whose base is the cross-section of the pump, and whose altitude is 1, by w; then will wh denote the pressure of the atmosphere exerted upwards through the water in the reservoir and pump.

Now, when the piston is at its lowest position, in order that it may not thrust open the piston valve and escape, the pressure of the confined air must be exactly equal to that of the external atmosphere; that is, equal to wh. When the piston is at its highest position, the confined air will be rarefied, the volume occupied being proportional to its height. Denoting the pressure of the rarefied air by wh', we shall have from MARIOTTE's law,

$$wh : wh' :: a - x : a - p - x.$$

$$\therefore wh' = wh \frac{a - p - x}{a - x}.$$

If the water does not rise when the piston is at its highest position, the pressure of the rarefied air, *plus* the weight of the column already raised, will be equal to the pressure of the external atmosphere; or

$$wh \frac{a - p - x}{a - x} + wx = wh.$$

Solving this equation with respect to x, we have,

$$x = \frac{a \pm \sqrt{a^2 - 4ph}}{2}.$$

If we have,

$$4ph > a^2; \quad \text{or,} \quad p > \frac{a^2}{4h},$$

the value of x will be imaginary, and there will be no point at which the water will cease to rise. Hence, the above inequality expresses the relation that must exist, in order that the pump may be effective. This condition expressed in words, gives the following rule:

The pump will be effective, when the play of the piston is greater than the square of the distance from the surface of the water in the reservoir, to the highest position of the piston, divided by four times the height at which the pressure of the atmosphere will support a column of water in a vacuum.

Let it be required to find the least allowable play of the piston, when the highest position of the piston is 16 feet above the water in the reservoir, and when the barometer stands at 28 inches.

In this case,

$a = 16$ ft., and $h = 28$ in. $\times 13\frac{1}{2} = 378$ in. $= 31\frac{1}{2}$ ft.

Hence,

$$p > \tfrac{256}{126} \text{ ft.}; \quad \text{or,} \quad p > 2\tfrac{2}{63} \text{ ft.}$$

To find the quantity of work required to make a double stroke of the piston, after the water reaches the level of the spout.

In depressing the piston, no force is required, except that necessary to overcome the inertia of the parts and the friction. Neglecting these for the present, the quantity of work in the downward stroke, may be regarded as 0. In raising the piston, its upper surface will be pressed downwards, by the pressure of the atmosphere wh, *plus* the weight of the column of water from the piston to the spout; and it will be pressed upwards, by the pressure of the atmosphere, transmitted through the pump, *minus* the weight of a column of water, whose cross-section is equal to that of the barrel, and whose altitude is the distance from the piston to the surface of the water in the reservoir. If we subtract the latter pressure from the former, the difference will be the resultant downward pressure. This difference will be equal to the weight of a column of water, whose base is the cross-section of the barrel, and whose height is the distance of the spout above the reservoir. Denoting the height by H, the pressure will be equal to wH. The path through which the pressure is exerted during the ascent of the piston, is equal to the play of the piston, or p. Denoting the quantity of work required, by Q, we shall have,

$$Q = wpH.$$

But wp is the weight of a volume of water, whose base is the cross-section of the barrel, and whose altitude is the play of the piston. Hence, the value of Q is equal to the quantity of work necessary to raise this volume of water from the level of the water in the reservoir to the spout. This volume is evidently equal to the volume actually delivered at each double stroke of the piston. Hence, the quantity of work expended in pumping with the sucking and lifting pump, all hurtful resistances being neglected, is equal to the quantity of work necessary to lift the amount of water, actually delivered, from the level of the water in the reservoir to the height of the spout. In addition to this work, a sufficient amount of power must be exerted, to overcome the hurtful resistances. The disadvantage of this pump, is the irregularity with which the force must act, being 0 in depressing the piston, and a maximum in raising it. This is an important objection when machinery is employed in pumping; but it may be either partially or entirely overcome, by using two pumps, so arranged, that the piston of one shall ascend as that of the other descends. Another objection to the use of this kind of pump, is the irregularity of flow, the inertia of the column of water having to be overcome at each upward stroke. This, by creating shocks, consumes a portion of the force applied.

Sucking and Forcing Pump.

210. This pump consists of a cylindrical barrel A, with its attached sucking-pipe B, and sleeping-valve G, as in the pump just discussed. The piston C is solid, and is worked up and down in the barrel by means of a lever F, attached to the piston-rod D. At the bottom of the barrel, a branch-pipe leads into an air-vessel K, through a second sleeping-valve F, which opens upwards, and closes by its own weight. A delivery-pipe H, enters the air-vessel at its top, and terminates near its bottom.

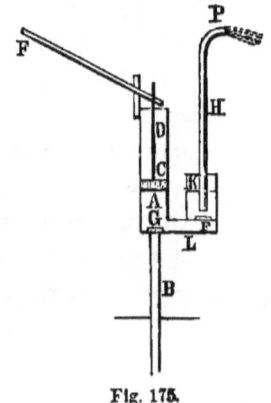

Fig. 175.

To explain the action of this

pump, suppose the piston C to be depressed to its lowest limit. Now, if the piston be raised to its highest position, the air in the barrel will be rarefied, its tension will be diminished, the air in the tube B, will thrust open the valve, and a portion of it will escape into the barrel. The pressure of the external air will then force a column of water up the pipe B, until the tension of the rarefied air, *plus* the weight of the column of water raised, is equal to the tension of the external air. An equilibrium being produced, the valve G closes by its own weight. If, now, the piston be again depressed, the air in the barrel will be condensed, its tension will increase till it becomes greater than that of the external air, when the valve F will be thrust open, and a portion of it will escape through the delivery-pipe H. After a few double strokes of the piston, the water will rise through the valve G, and then, as the piston descends, it will be forced into the air-vessel, the air will be condensed in the upper part of the vessel, and, acting by its elastic force, will force a portion of the water up the delivery-pipe and out at the spout P. The object of the air-vessel is, to keep up a continued stream through the pipe H, otherwise it would be necessary to overcome the inertia of the entire column of water in the pipe at every double stroke. The flow having commenced, at each double stroke, a volume of water will be delivered from the spout, equal to that of a cylinder whose base is the area of the piston, and whose altitude is the play of the piston.

The same relative conditions between the parts should exist as in the sucking and lifting pump.

To find the quantity of work consumed at each double stroke, after the flow has become regular, hurtful resistances being neglected:

When the piston is descending, it is pressed downwards by the tension of the air on its upper surface, and upwards by the tension of the atmosphere, transmitted through the delivery-pipe, *plus* the weight of a column of water whose base is the area of the piston, and whose altitude is the distance of the spout above the piston. This distance is variable during the stroke, but its mean value is the distance of the middle of the play below the spout. The difference between these pressures is exerted upwards, and is equal to the weight of a column of water whose base is the area of the piston, and whose altitude is the distance from the middle of the play to the spout. The distance through which the force is exerted, is equal to the play of the piston. Denoting the quantity of work during the descending stroke, by Q'; the weight of a column of water, having a base equal to the area of the piston, and a unit in altitude, by w; and the height of the spout above the middle of the the play, by h', we shall have,

$$Q' = wh' \times p.$$

When the piston is ascending, it is pressed downwards by the tension of the atmosphere on its upper surface, and upwards by the tension of the atmosphere, transmitted through the water in the reservoir and pump, *minus* the weight of a column of water whose base is the area of the piston, and whose altitude is the height of the piston above the reservoir. This height is variable, but its mean value is the height of the middle of the play above the water in the reservoir. The distance through which this force is exerted, is equal to the play of the piston. Denoting the quantity of work during the ascending stroke, by Q'', and the height of the middle of the play above the reservoir, by h'', we have,

$$Q'' = wh'' \times p.$$

Denoting the entire quantity of work during a double stroke, by Q, we have,

$$Q = Q' + Q'' = wp(h' + h'').$$

But wp is the weight of a volume of water, the area of whose base is that of the piston, and whose altitude is the play of the piston; that is, it is the weight of the volume delivered at the spout at each double stroke.

The quantity $h' + h''$, is the entire height of the spout above the level of the cistern. Hence, the quantity of work expended, is equal to that required to raise the entire volume delivered, from the level of the water in the reservoir to the height of the spout. To this must be added the work necessary to overcome the hurtful resistances, such as friction, &c.

If $h' = h''$, we shall have, $Q' = Q''$; that is, the quantity of work during the ascending stroke, will be equal to that during the descending stroke. Hence, the work of the motor will be more nearly uniform, when the middle of the play of the piston is at equal distances from the reservoir and spout.

Fire Engine.

211. The fire engine is essentially a double sucking and forcing pump, the two piston rods being so connected, that when one piston ascends the other descends. The sucking and delivery pipes are made of some flexible material, generally of leather, and are attached to the machine by means of metallic screw joints.

The figure exhibits a cross-section of the essential part of a Fire Engine.

A A' are the two barrels, C C' the two pistons, connected by the rods, DD, with the lever, E E'. B is the sucking pipe, terminating in a box from which the water may enter either barrel through the valves, G G'. K is the air vessel, common to both pumps, and communicating with them by the valves F F'. H is the delivery pipe.

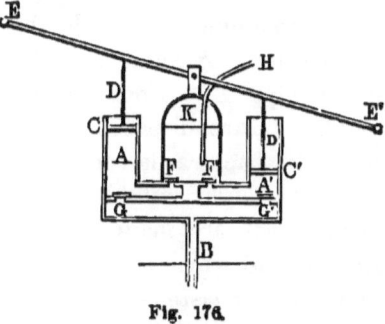

Fig. 176.

The instrument is mounted on wheels for convenience of transportation. The lever EE' is worked by means of rods at right angles to the lever, so arranged that several men can apply their strength in working the pump. The action of the pump differs in no respect from that of the forcing pump; but when the instrument is worked vigorously, there is more water forced into the air vessel, the tension of the air is very much augmented, and its elastic force, thus brought into play, propels the water to a considerable distance from the mouth of the delivery pipe. It is this capacity of throwing a jet of water to a great distance, that gives to the engine its value in extinguishing fires.

A pump entirely similar to the fire engine in its construction, is often used under the name of the double action forcing pump for raising water for other purposes.

The Rotary Pump.

212. The rotary pump is a modification of the sucking and forcing pump. Its construction will be best understood from the drawing, which represents a vertical section through the axis of the sucking-pipe, and at right angles to axis of the rotary portion of the pump.

A represents an annular ring of metal, which may be made to revolve about its axis O. DD is a second ring of metal, concentric with the first, and forming with it an intermediate annular space. This space communicates with the sucking-pipe K, and the delivery pipe L. Four radial paddles C, are disposed so as to slide backwards and forwards through suitable openings, which are made in the ring A, and which are moved around with it.

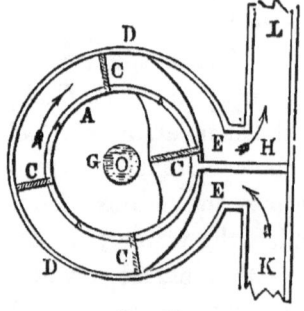

Fig. 177.

G is a solid guide, firmly fastened to the end of the cylinder enclosing the rotary apparatus, and cut as represented in the figure. $E\,E$ are two springs, attached to the ring D, and acting by their elastic force, to press the paddles firmly against the guide. These springs are of such dimensions as not to impede the flow of the water *from* the pipe K, and *into* the pipe L.

When the axis O is made to revolve, each paddle, as it reaches and passes the partition H, is pressed against the guide, but, as it moves on, it is forced, by the form of the guide, against the outer wall D. The paddle then drives the air in front of it, around, in the direction of the arrow-head, and finally expels it through the pipe L. The air behind the paddle is rarefied, and the pressure of the external air forces a column of water up the pipe. As the paddle approaches the opening to the pipe L, the paddle is pressed back by the spring E, against the guide, and an outlet into the ascending pipe L, is thus provided. After a few revolutions, the air is entirely exhausted from the pipe K. The water enters the channel $C\,C$, and is forced up the pipe L, from which it escapes by a spout at the top. The quantity of work expended in raising a volume of water to the spout, by this pump, is equal to that required to lift it through the distance from the level of the water in the cistern to the spout. This may be shown in the same manner as was explained under the head of the sucking and forcing-pump. To this quantity of work, must be added the work necessary to overcome the hurtful resistances, as friction, &c.

This pump is well adapted to machine pumping, the work being very nearly uniform.

A machine, entirely similar to the rotary pump, might be constructed for exhausting foul air from mines; or, by reversing the direction of rotation, it might be made to force a supply of fresh air to the bottom of deep mines.

Besides the pumps already described, a great variety of others have been invented and used. All, however, depend upon some modification of the principles that have just been discussed.

The Hydrostatic Press.

213. The hydrostatic press is a machine for exerting great pressure through small spaces. It is much used in compressing seeds to obtain oil, in packing hay and bales of goods, also in raising great weights. Its construction, though requiring the use of a sucking-pump, depends upon the principle of equal pressures (Art. 154).

It consists essentially of two vertical cylinders, A and B, each provided with a solid piston. The cylinders communicate by means of a pipe C, whose entrance to the larger cylinder is closed by a sleeping valve E. The smaller cylinder communicates with the reservoir of water K, by a sucking-pipe H, whose upper extremity is closed by the sleeping-valve D.

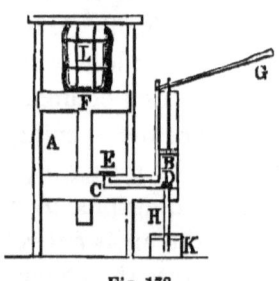

Fig. 178.

The smaller piston B, is worked up and down by the lever G. By working the lever G, up and down, the water is raised from the reservoir and forced into the larger cylinder A; and when the space below the piston F is filled, a force of compression is exerted upwards, which is as many times greater than that applied to the piston B, as the area of F is greater than B (Art. 154). This force may be utilized in compressing a body L, placed between the piston and the frame of the press.

Denote the area of the larger piston by P, of the smaller, by p, the pressure applied to B, by f, and that exerted at F, by F; we shall have,

$$F : f :: P : p, \quad \therefore \; F = \frac{fP}{p}.$$

If we denote the longer arm of the lever G, by L, and

the shorter arm, by l, and represent the force applied at the extremity of the longer arm, by K, we shall have from the principle of the lever (Art. 78),

$$K : f :: l : L, \quad \therefore f = \frac{KL}{l}.$$

Substituting this value of f above, we have,

$$F = \frac{PKL}{pl}.$$

To illustrate, let the area of the larger piston be 100 square inches, that of the smaller piston 1 square inch; suppose the longer arm of the lever to be 30 inches, and the shorter arm to be 2 inches, and a force of 100 pounds to be applied at the end of the longer arm of the lever; to find the pressure exerted upon F.

From the conditions,

$$P = 100, \quad K = 100, \quad L = 30, \quad p = 1, \text{ and } l = 2.$$

Hence,

$$F = \frac{100 \times 100 \times 30}{2} = 150000 \text{ lbs.}$$

We have not taken into account the hurtful resistances, hence, the total pressure of 150000 pounds must be somewhat diminished.

The volume of water forced from the smaller to the larger piston, during a single descent of the piston F'', will occupy in the two cylinders, spaces whose heights are inversely as the areas of the pistons. Hence, the path, over which f is exerted, is to the path over which F is exerted, as P is to p. Or, denoting these paths by s and S, we have,

$$s : S :: P : p;$$

or, since $P : p :: F : f$, we shall have,

$$s : S :: F : f, \quad \therefore fs = FS.$$

326 MECHANICS.

That is, *the quantities of work of the power and resistance are equal,* a principle which holds good in all machines.

EXAMPLES.

1. The cross-section of a sucking and forcing pump is 6 square feet, the play of the piston 3 feet, and the height of the spout, above the level of the reservoir, 50 feet. What must be the effective horse power of an engine which can impart 30 double strokes per minute, hurtful resistances being neglected?

SOLUTION.

The number of units of work required to be performed each minute, is equal to

$$6 \times 3 \times 50 \times 62\tfrac{1}{2} = 56250.$$

Hence,
$$n = \tfrac{56250}{33000} = 1\tfrac{93}{132}. \quad Ans.$$

2. In a hydrostatic press, the areas of the two pistons are, respectively, 2 and 400 square inches, and the two arms of the lever are, respectively, 1 and 20 inches. Required the pressure on the larger piston for each pound of pressure applied to the longer arm of the lever?   *Ans.* 4000 lbs.

3. The areas of the two pistons of a hydrostatic press are, respectively, equal to 3 and 300 square inches, and the shorter arm of the lever is one inch. What must be the length of the longer arm, that a force of 1 lb. may produce a pressure of 1000 lbs.   *Ans.* 10 inches.

The Siphon.

214. The siphon is a bent tube, used for transferring a liquid from a higher to a lower level, over an intermediate elevation The siphon consists of two branches, *AB* and *BC*, of which the outer one is the longer. To use the instrument, the tube is filled with the liquid in any manner, the end of the longer branch being stopped with the finger or a stop-cock, in which case, the pressure of the atmosphere will prevent the liquid from escaping

Fig. 179.

at the other end. The instrument is then inverted, the end C being submerged in the liquid, and the stop removed from A. The liquid will begin to flow through the tube, and the flow will continue till the level of the liquid in the reservoir reaches that of the mouth of the tube C.

To find the velocity with which water will issue from the siphon, let us consider an infinitely small layer at the orifice A. This layer will be pressed downwards, by the tension of the atmosphere exerted on the surface of the reservoir, diminished by the weight of the water in the branch BD, and increased by the weight of the water in the branch BA. It will be pressed upwards by the tension of the atmosphere acting directly upon the layer. The difference of these forces, is the weight of the water in the portion of the tube DA, and the velocity of the stratum will be due to that weight. Denoting the vertical height of DA, by h, we shall have, for the velocity (Art. 173),

$$v = \sqrt{2gh}.$$

This is the theoretical velocity, but it is never quite realized in practice, on account of resistances, which have been neglected in the preceding investigation.

The siphon may be filled by applying the mouth to the end A, and exhausting the air by suction. The tension of the atmosphere, on the upper surface of the reservoir, will press the water up the tube, and fill it, after which the flow will go on as before. Sometimes, a sucking-tube AD, is inserted near the opening A, and rising nearly to the bend of the siphon. In this case, the opening A, is closed, and the air exhausted through the sucking-tube AD, after which the flow goes on as before.

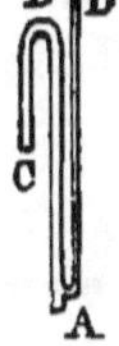

Fig. 186.

The Wurtemburg Siphon.

215. In the Wurtemburg siphon, the ends of the tube are

bent twice, at right-angles, as shown in the figure. The advantage of this arrangement is, that the tube, once filled, remains so, as long as the plane of its axis is kept vertical. The siphon may be lifted out and replaced at pleasure, thereby stopping the flow at will.

Fig. 181.

It is to be observed that the siphon is only effectual when the distance from the highest point of the tube to the level of the water in the reservoir is less than the height at which the atmospheric pressure will sustain a column of water in a vacuum. This will, in general, be less than 34 feet.

The Intermitting Siphon.

216. The intermitting siphon is represented in the figure. AB is a curved tube issuing from the bottom of a reservoir. The reservoir is supplied with water by a tube E, having a smaller bore than that of the siphon. To explain its action, suppose the reservoir at first to be empty, and the tube E to be opened; as soon as the reservoir is filled to the level of CD, the water will begin to flow from the opening

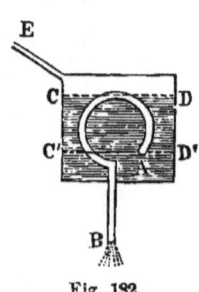

Fig. 182.

B, and the flow once commenced, will continue till the level of the reservoir is again reduced to the level $C'D'$, drawn through the opening A. The flow will then cease till the cistern is again filled to CD, and so on as before.

Intermitting Springs.

217. Let A represent a subterranean cavity, communicating with the surface of the earth by a channel ABC, bent like a siphon. Suppose the reservoir to be fed by percolation through the crevices, or by a small channel D. When the

Fig. 183.

water in the reservoir rises to the height of the horizontal plane BD, the flow will commence at C, and, if the channel is sufficiently large, the flow will continue till the water is reduced to the level plane drawn through C. An intermission of flow will occur till the reservoir is again filled, and so on, intermittingly. This phenomena has been observed at various places.

Siphon of Constant Flow.

218. We have seen that the velocity of efflux depends upon the height of the water in the reservoir above the external opening of the siphon. When the water is drawn off from the reservoir, the upper surface sinks, this height diminishes, and, consequently, the velocity continually diminishes.

If, however, the shorter branch CD, of the tube, be inserted through a piece of cork large enough to float the siphon, the instrument will sink as the upper surface is depressed, the height of DA will remain the same, and, consequently, the flow will be uniform till the bend of the siphon comes in contact with the upper edge of the reservoir. By suitably adjusting the siphon in the cork, the velocity of efflux can be increased or decreased within certain limits. In this manner, any desired quantity of the fluid can be drawn off in a given time.

The siphon is used in the arts, for decanting liquids, when it is desirable not to stir the sediment at the bottom of a vessel. It is also employed to draw a portion of a liquid from the interior of a vessel when that liquid is overlaid by one of less specific gravity.

The Hydraulic Ram.

219. The hydraulic ram is a machine for raising water by means of shocks caused by the sudden stoppages of a stream of water.

The instrument consists of a reservoir B, which is supplied with water by an inclined pipe A; on the upper surface of the reservoir, is an orifice which may be closed by a spherical valve *D*; this valve, when not pressed against the opening, rests in a metallic framework immediately below the orifice; *G* is an air-vessel communicating with the reservoir by an orifice *F*, which is fitted with a spherical valve *E*; this valve closes the orifice *F*, except when forced upwards, in which case its motion is restrained by a metallic frame work or cage; *H* represents a delivery-pipe entering the air-vessel at its upper part, and terminating near the bottom. At *P* is a small valve, opening inwards, to supply the loss of air in the air-vessel, arising from absorption by the water in passing through the air vessel.

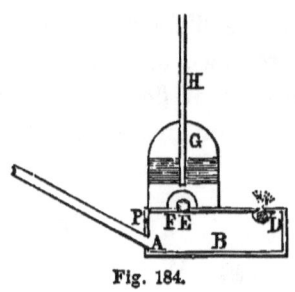

Fig. 184.

To explain the action of the instrument, suppose, at first, that it is empty, and all the parts in equilibrium. If a current of water be admitted to the reservoir, through the inclined pipe *A*, the reservoir will soon be filled, and commence rushing out at the orifice *C*. The impulse of the water will force the spherical valve *D*, upwards, closing the opening; the velocity of the water in the reservoir will be suddenly checked; the reaction will force open the valve *E*, and a portion of the water will enter the air-chamber *G*. The force of the shock having been expended, the spherical valves will both fall by their own weight; a second shock will take place, as before; an additional quantity of water will be forced into the air-vessel, and so on, indefinitely. As the water is forced up into the air-vessel, the air becomes compressed; and acting by its elastic force, it urges a stream of water up the pipe *H*. The shocks occur in rapid succession, and, at each shock, a quantity of water is forced into the air-chamber, and thus a constant stream is kept up. To explain the use of the valve *P*, it may be remarked that water absorbs more air under a great pressure, than under a smaller one. Hence, as it passes through the air-chamber, a portion of the air contained is taken up by the water and carried out through the pipe H. But each time that the valve D falls, there is a tendency to produce a vacuum in the upper part of the reservoir, in consequence of the rush of the fluid to escape through the opening. The pressure of the external air then forces the valve P open, a small portion of air enters, and is afterwards forced up with the water into the vessel G, to keep up the supply.

The hydraulic ram is only used where it is required to raise small quantities of water, such as for the supply of a house, or garden. Only a small fraction of the amount of fluid which enters the supply-pipe actually passes out through the delivery-pipe; but, if the head of water is pretty large, the column may be raised to a great height. Water is often raised, in this manner, to the highest points of lofty buildings.

Sometimes, an additional air-vessel is introduced over the valve E, for the purpose of deadening the shock of the valve in its play up and down.

Archimedes' Screw.

220. This machine is intended for raising water through small heights, and consists, in its simplest form, of a tube wound spirally around a cylinder. This cylinder is mounted so that its axis is oblique to the horizon, the lower end dipping into the reservoir. When the cylinder is turned on its axis, by a crank attached to its upper extremity, the lower end of the tube describes a circumference of a circle, whose plane is perpendicular to the axis. When the mouth of the tube comes to the level of the axis and begins to ascend, there will be a certain quantity of water in the tube, which will flow so as to occupy the lowest part of the spire; and, if the cylinder is properly inclined to the horizon, this flow will be towards the upper end of the tube. At each revolution, an additional quantity of water will enter the tube, and that already in the tube will be forced, or raised, higher and higher, till, at last, it will flow from the orifice at the upper end of the spiral tube.

The Chain Pump.

221. The chain pump is an instrument for raising water through small elevations. It consists of an endless chain passing over two wheels, A and B, having their axes horizontal, the one being below the surface of the water, and the other above the spout of the pump. Attached to this chain, and at right angles to it, are a system of circular disks, just fitting the tube CD. If the cylinder A be turned in the direction of the arrow-head, the buckets or disks will rise through the tube

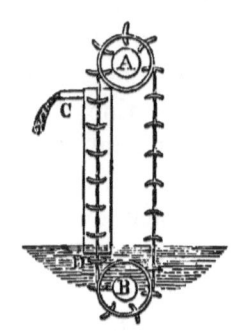

Fig. 185.

CD, carrying the water in the tube before them, until it reaches the spout C, and escapes. The buckets thus emptied return through the air to the reservoir, and so on perpetually. One great objection to this machine is, the difficulty of making the buckets fit the tube of the pump. Hence there is a constant leakage, requiring a great additional expenditure of force.

Sometimes, instead of having the body of the pump vertical, it is inclined; in which case it does not differ much in principle from the wheel with flat buckets, that has been used for raising water.

The Air Pump.

222. The air pump is a machine for rarefying the air in a closed space.

It consists of a cylindrical barrel A, in which a piston B, fitting air-tight, is worked up and down by a lever C, attached to a piston-rod D. The barrel communicates with an air-tight ves-

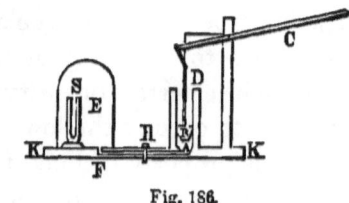

Fig. 186.

sel E, called a receiver, by means of a narrow pipe. The receiver, which is usually of glass, is ground so as to fit air-tight upon a smooth bed-plate KK. The joint between the receiver and plate may be rendered more perfectly air-tight by rubbing it with a little oil. A stop-cock H, of a peculiar construction, permits communication to be made at pleasure between the barrel and receiver, or between the barrel and the external air. When the stop-cock is turned in a particular direction, the barrel and receiver are made to communicate; but on turning it through 90 degrees, the communication with the receiver is cut off, and a communication is opened between the barrel and the external air. Instead of the stop-cock, valves are often used, which are either opened and closed by the elastic force of the air, or by the force that works the pump. The communicating pipe should be exceedingly small, and the piston B should, when at its lowest point, fit accurately to the bottom of the barrel.

To explain the action of the air pump, suppose the piston to be depressed to its lowest position. The stop-cock H, is turned so as to open a communication between the barrel and receiver, and the piston is raised to its highest point by a force applied to the lever C. The air which before occupied the receiver and pipe, will expand so as to fill the barrel, receiver, and pipe. The stop-cock is then turned so as to cut off communication between the barrel and receiver, and open the barrel to the external air, and the piston again depressed to its lowest position. The rarefied air in the barrel is expelled into the external air by the depression of the piston. The air in the receiver is now more rarefied than at the beginning, and by a continued repetition of the process just described, any degree of rarefaction may be attained.

To measure the degree of rarefaction of the air in the receiver, a siphon-gauge may be used, or a glass tube, 30 inches long, may be made to communicate at its upper extremity with the receiver, whilst its lower extremity dips into a cistern of mercury. As the air is rarefied in the receiver, the pressure on the mercury in the tube becomes less than that on the surface of the mercury in the cistern, and the mercury rises in the tube. The tension of the air in the receiver will be given by the difference between the height of the barometric column and that of the mercury in the tube.

To investigate a formula for computing the tension of the air in the receiver, after any number of double strokes, let us denote the capacity of the receiver in cubic feet, by r, that of the connecting-pipe, by p, and the space between the bottom of the barrel and the highest position of the piston, by b. Denote the original tension of the air, by t; its tension after the first upward stroke of the piston, by t'; after the second, third, ... n^{th}, upward strokes, by t', t'', ... $t^{n'}$.

The air which originally occupied the receiver and pipe, fills the receiver, pipe, and barrel, after the first upward stroke; according to MARIOTTE's law, its tension in the two cases varies inversely as the volumes occupied; hence,

$$t \cdot t' :: p + r + b : p + r, \quad \therefore \quad t' = t \frac{p+r}{p+r+b}.$$

In like manner, we shall have, after the second upward stroke,

$$t' : t'' :: p + r + b : p + r, \quad \therefore \quad t'' = t' \frac{p+r}{p+b+r}.$$

Substituting for t' its value, deduced from the preceding equation, we have,

$$t'' = t \left(\frac{p+r}{p+b+r} \right)^2.$$

In like manner, we find,

and, in general,
$$t_{n'} = t\left(\frac{p+r}{p+b+r}\right)^n.$$

If the pipe is exceedingly small, its capacity may be neglected in comparison with that of the receiver, and we shall then have,
$$t_{n'} = t\left(\frac{r}{b+r}\right)^n.$$

Let it be required, for example, to determine the tension of the air after 5 upward strokes, when the capacity of the barrel is one-third that of the receiver.

In this case, $\frac{r}{b+r} = \frac{3}{4}$, and $n = 5$, whence,

$$t^v = t\tfrac{243}{1024}.$$

Hence, the tension is less than a fourth part of that the external air.

Instead of the receiver, the pipe may be connected by a screw-joint with any closed vessel, as a hollow globe or glass flask. In this case, by reversing the direction of the stop-cock, in the up and down motion of the piston, the instrument may be used as a condenser. When so used, the tension, after n downward strokes of the piston, is given by the formula,
$$t_{n'} = t\left(\frac{b+r}{r}\right)^n.$$

Taking the same case as that before considered, with the exception that the instrument is used as a condenser instead of a rarefier, we have, after 5 downward strokes,

$$t^v = t\tfrac{1024}{243}.$$

That is, the tension is more than four times that of the external air.

When the pump is used for condensing air, it is called a *condenser*.

Artificial Fountains.

223. An artificial fountain is an instrument by means of which a liquid is forced upwards in the form of a jet, by the tension of condensed air. The simplest form of an artificial fountain is called HERO's ball.

Hero's Ball.

224. This instrument consists of a hollow globe A, into the top of which is inserted a vertical tube B, reaching nearly to the bottom of the globe. This tube is provided with a stop-cock C, by means of which it may be closed, or opened to the external air, at pleasure. A second tube D, enters the globe near the top, which is also provided with a stop-cock E.

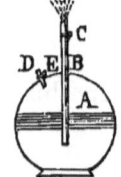

Fig. 187.

To use the instrument, close the stop-cock C, and fill the lower portion of the globe with water through the tube D; then attach the tube D to a condenser, and pump air into the upper part of the globe, and confine it there by closing the stop-cock E. If, now, the stop-cock C be opened, the pressure of the confined air on the surface of the water in the globe, will force a jet up through the tube B. This jet will rise to a greater or less height, according to the greater or less quantity of air that was forced into the globe. The water will continue to flow through the tube as long as the tension of the confined air is greater than that of the external atmosphere, or else till the level of the water in the globe reaches the lower end of the tube.

Instead of using the condenser, air may be introduced by blowing with the mouth through the tube D, and then confined as before, by turning the stop-cock E.

The principle of HERO's ball is the same as that of the air-chamber in the forcing pump and fire-engine, already explained.

Hero's Fountain.

225. Hero's fountain is constructed on the same principle as Hero's ball, except that the compression of the air is effected by the weight of a column of water, instead of by aid of a condenser.

A represents a cistern, similar to Hero's ball, with a tube *B*, extending nearly to the bottom of the cistern. *C* is a second cistern placed at some distance below *A*. This cistern is connected with a basin *D*, by a bent tube *E*, and also with the upper part of the cistern *A*, by a tube *F*. When the fountain is to be used, the cistern *A* is nearly filled with water, the cistern *C* being empty. A quantity of water is then poured into the basin *D*, which, acting by its weight, sinks into the cistern *C*, compressing the air in the upper portion of it into a smaller space, thus increasing its tension. This increase of tension acting on the surface of the water in *A*, forces a jet through the tube *B*, which rises to a greater or less height according to the greater or less increase of the atmospheric tension. The flow will continue till the level of the water in *A*, reaches the bottom of the tube *B*. The measure of the compressing force on a unit of surface of the water in *C*, is the weight of a column of water, whose base is a square unit, and whose altitude is the difference of level between the water in *D* and *C*.

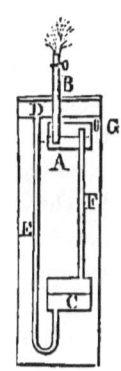

Fig. 183

If Hero's ball be partially filled with water and placed under the receiver of an air pump, the water will be observed to rise in the tube, forming a fountain, as the air in the receiver is exhausted. The principle is the same as before, an excess of pressure on the water within the globe over that without. In both cases, the flow is resisted by the tension of the air without, and is urged on by the tension within.

Wine-Taster and Dropping-Bottle.

226. The wine-taster is used to bring up a small portion of wine or other liquid, from a cask. It consists of a tube, open at the top, and terminating below in a very narrow tube, also open. When it is to be used, it is inserted to any depth in the liquid, which will rise in the tube to the level of the upper surface of that liquid. The finger is then placed so as to close the upper orifice of the tube, and the instrument is raised out of the cask. A portion of the fluid escapes from the lower orifice, until the pressure of the rarefied air in the tube, plus the weight of a column of liquid, whose cross-section is that of the tube, and whose altitude is that of the column of fluid retained, is just equal to the pressure of the external air. If the tube be placed over a tumbler, and the finger removed from the upper orifice, the fluid brought up will escape into the tumbler.

Fig. 189.

If the lower orifice is very small, a few drops may be allowed to escape, by taking off the finger and immediately replacing it. The instrument then constitutes the dropping tube.

The Atmospheric Inkstand.

227. The atmospheric inkstand consists of a cylinder A, which communicates by a tube with a second cylinder B. A piston C, is moved up and down in A, by means of a screw D. Suppose the spaces A and B, to be filled with ink. If the piston C is raised, the pressure of the external air forces the ink to follow it, and the part B is emptied. If the operation be reversed, and the piston C depressed, the ink is again forced into the space B. This operation may be repeated at pleasure.

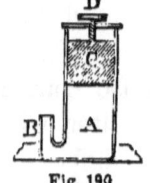

Fig. 190.

APPENDIX.

The following notes contain elementary demonstrations of those principles, which in the body of the work are proved by means of the Calculus.

Note on Articles 64—70; pp. 72–76.

These articles may be omitted without at all impairing the unity of the subject, the preceding principles being sufficient to find the centre of gravity of all bodies, approximatively.

Note on Articles 112—114; pp. 143—148.

The principal formulas in these articles may be deduced as follows:

112. By definition, a body moves uniformly when it passes over equal spaces in equal times; now if it passes over a space v in one second, it will pass over t times that space in t seconds; that is, it will pass over a space vt. If we suppose it to have passed over a space s' before the commencement of the time t, we shall have for the entire space passed over, and which may be denoted by s,

$$s = vt + s' \quad . \quad . \quad . \quad . \quad (58.)$$

This equation corresponds to Equation (58) of the text.

113. The formulas of Article 113 may be omitted without impairing the unity of the course. They are only of use in Higher Mechanics, where the employment of the Calculus is a necessity.

114. *Uniformly varied motion*, is that in which the velocity increases or diminishes uniformly. In the former case the motion is *accelerated*, in the latter it is *retarded*. In both cases the moving force is *constant*.

Denote the moving force by f, the mass moved being the unit of mass.

According to Art. 24, the measure of the force is the velocity impressed in a unit of time, that is, in 1 second. Now from the principal of inertia, Art. 18, it follows that a force will produce the same general effect upon a body, whether it finds the body at rest or in motion. Hence, the velocity impressed in any second of time is constant; that is, if the velocity impressed in one second of time is f, in t seconds it will be t times f, or ft. Denoting the velocity by v, we shall have,

$$v = ft \quad \ldots \quad \quad (69.)$$

If the body has a velocity v' at the beginning of the time t, this velocity is called the initial velocity. Adding this to the velocity imparted during the time t, we have,

$$v = v' + ft \quad \ldots \quad (67.)$$

With respect to the space passed over, it may be remarked that the velocity increases uniformly; hence the space passed over in any time, is the same that it would have passed over in the same time, had it moved uniformly during that time with its mean or *average* velocity. Now, if a body start from a state of rest, its velocity at starting is 0, and at the end of the time t it is ft, Equation (69); the average or mean of these is $\frac{1}{2} ft$. But the space described in the time t, when the body moves with the uniform velocity $\frac{1}{2} ft$, is (Equation 55) equal to $\frac{1}{2} ft \times t$; denoting the space by s, we have,

$$s = \tfrac{1}{2} f t^2 \quad \ldots \quad (70.)$$

If in Equation (7), we make $t = 1$, we have,

$$s = \tfrac{1}{2} f; \quad \text{or,} \quad f = 2s;$$

that is, if a body moves from a state of rest, the space described in the first second of time, is equal to half the measure of the accelerating force; or, the acceleration is measured by twice the space passed over in one second of time.

If we suppose that a body starts from rest before the beginning of the time t, so as to pass over a space s' before the beginning of t, it will during that time have acquired some velocity, which we may denote by v'. The space reckoned from the origin of spaces up to the position of the body at the end of the time t, is made up of three parts; first, the space s', called the *initial* space; second, a space due to the velocity v' during the time t, which is measured by $v't$; third, a space due to the action of the incessant force during the time t, which will (Equation 70) be equal to $\frac{1}{2}ft^2$. Adding these together, we have finally,

$$s = s' + v't + \tfrac{1}{2}ft^2 \quad . \quad . \quad . \quad (68.)$$

If, in Equations (67) and (68), we suppose f to be essentially positive, the motion will be *accelerated;* if we suppose it to be essentially negative, the motion will be *retarded*, and these equations become

$$v = v' - ft \quad . \quad . \quad . \quad . \quad . \quad (71.)$$

$$s = s' + v't - \tfrac{1}{2}ft^2 . \quad . \quad . \quad (72.)$$

Note on Article 121, pp. 163—164.

The formula deduced in the first part of this article is needed in the investigations of Acoustics and Optics, and can only be found by the Calculus. This part of the article may be omitted without impairing the unity of the course.

Note on Article 123, pp. 166—168.

This article, up to the end of Equation (95), may be replaced by the following demonstration:

The simple pendulum.

123. A PENDULUM is a heavy body suspended from a horizontal axis about which it is free to vibrate.

In order to investigate the circumstances of vibration, let us first consider the hypothetical case of a single material point, vibrating about an axis to which it is attached by a rod destitute of weight. Such a pendulum is called a SIMPLE PENDULUM. The laws of vibration in this case will be identical with those explained in Art. 120, the arc ABC being an arc of a circle.

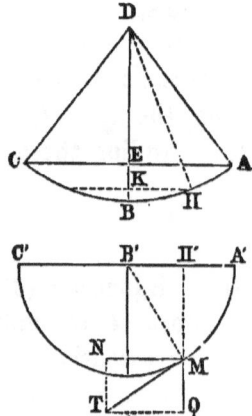

Let ABC be the arc through which the vibration takes place, and denote its radius DA, by l. The angle ADC is called the *amplitude* of vibration; half of this angle, ADB, is called the *angle of deviation*.

If the point starts from rest at A, it will, on reaching any point H, have a velocity v, due to the height EK, denoted by h, (Art. 120). Hence,

$$v = \sqrt{2gh} \quad \ldots \quad (92.)$$

Let us suppose that the angle of deviation is so small, that the chords of the arcs AB and HB, may be considered equal to the arcs themselves. We shall have (Davies' Legendre, Bk. IV., Prop. XXIII., Cor.),

$$\overline{AB}^2 = 2l \times EB, \quad \text{and} \quad \overline{HB}^2 = 2l \times KB,$$

whence, by subtraction,

$$\overline{AB}^2 - \overline{HB}^2 = 2l(EB - KB) = 2l \times h.$$

Denoting AB by a, and HB by x, and solving the last equation, we have,

$$h = \frac{a^2 - x^2}{2l}.$$

Substituting this value of h in (92) it becomes,

$$v = \sqrt{\frac{g}{l}(a^2 - x^2)} \quad \ldots \quad (a.)$$

Now let us develop the arc ABC into a straight line $A'B'C'$, and suppose a material point to start from A' at the same time that the pendulum starts from A, and to vibrate back and forth upon $A'B'C'$ with the same velocities as the pendulum; then, when the pendulum is at any point H, this material point will be at the corresponding point H', and the times of vibration of the two will be exactly the same.

To find the time of vibration along the line $A'B'C'$, describe upon it a semi-circle $A'MC'$, and suppose a third material point to start from A' at the same time as the second, and to move uniformly around the arc with a velocity equal to $a\sqrt{\frac{g}{l}}$. Then will the time required for this particle to reach C' be equal to the space divided by the velocity (Art. 112). Denoting this time by t, and remembering that $A'B' = a$, we shall have,

$$t = \frac{\pi a}{a\sqrt{\frac{g}{l}}} = \pi\sqrt{\frac{l}{g}}.$$

Make $HB' = x$, and draw $H'M$ perpendicular to $A'C'$, and at M decompose the velocity of the third particle MT into two components MN and MQ, respectively parallel and perpendicular to $A'C'$.

We shall have for the horizontal component MN,

$$MN = MT \cos TMN.$$

But, $MT = a\sqrt{\dfrac{g}{l}}$, and because MT and MN are respectively perpendicular to $B'M$ and $H'M$, we have, $\cos TMN = \cos B'MH' = \dfrac{H'M}{B'M}$. But $B'M = a$, and $H'M = \sqrt{a^2 - x^2}$; hence, $\cos TMN = \dfrac{\sqrt{a^2-x^2}}{a}$. Substituting these values in Equation (b), we have for the horizontal velocity,

$$MN = \sqrt{\dfrac{g}{l}(a^2 - x^2)},$$

which is the same value as that obtained for v in Equation (a). Hence, we infer that the velocity of the third material point in the direction of $A'C'$ is always equal to that of the second point, consequently the times required to pass from A' to C' must be equal; that is, the time of vibration of the second point, and consequently of the pendulum, must be $\pi\sqrt{\dfrac{l}{g}}$. Denoting this time by t, we have,

$$t = \pi\sqrt{\dfrac{l}{g}} \quad . \quad . \quad . \quad . \quad (95.)$$

Note on Article 131, pp. 182—186.

This article may be omitted without impairing the unity of the course. The results may be assumed if needed. They can only be deduced by the Calculus by demonstrations too tedious for an Elementary Course.

THE NATIONAL SERIES OF STANDARD SCHOOL-BOOKS.

MATHEMATICS.

DAVIES'S COMPLETE SERIES.

ARITHMETIC.
Davies' Primary Arithmetic.
Davies' Intellectual Arithmetic.
Davies' Elements of Written Arithmetic.
Davies' Practical Arithmetic.
Davies' University Arithmetic.

TWO-BOOK SERIES.
First Book in Arithmetic, Primary and Mental.
Complete Arithmetic.

ALGEBRA.
Davies' New Elementary Algebra.
Davies' University Algebra.
Davies' New Bourdon's Algebra.

GEOMETRY.
Davies' Elementary Geometry and Trigonometry.
Davies' Legendre's Geometry.
Davies' Analytical Geometry and Calculus.
Davies' Descriptive Geometry.
Davies' New Calculus.

MENSURATION.
Davies' Practical Mathematics and Mensuration.
Davies' Elements of Surveying.
Davies' Shades, Shadows, and Perspective.

MATHEMATICAL SCIENCE.
Davies' Grammar of Arithmetic.
Davies' Outlines of Mathematical Science.
Davies' Nature and Utility of Mathematics.
Davies' Metric System.
Davies & Peck's Dictionary of Mathematics.

NATURAL SCIENCE— *Continued.*

THE NEW SURVEYING.

Van Amringe's Davies' Surveying.

By Charles Davies, LL.D., author of a Full Course of Mathematics. Revised by J Howard Van Amringe, A.M., Ph.D., Professor of Mathematics in Columbia College 566 pages. 8vo. Full sheep.

Davies' Surveying originally appeared as a text-book for the use of the United States Military Academy at West Point. It proved acceptable to a much wider field, and underwent changes and improvements, until the author's final revision, and has remained the standard work on the subject for many years.

In the present edition, 1883, while the admirable features which have hitherto commended the work so highly to institutions of learning and to practical surveyors have been retained, some of the topics have been abridged in treatment, and some enlarged. Others have been added, and the whole has been arranged in the order of progressive development. A change which must prove particularly acceptable is the transformation of the article on mining-surveying into a complete treatise, in which the location of claims on the surface, the latest and best methods of underground traversing, &c., the calculation of ore-reserves, and all that pertains to the work of the mining-surveyor, are fully explained and illustrated by practical examples. Immediately on the publication of this edition it was loudly welcomed in all quarters. A letter received as we write, from Prof. R. C. Carpenter, of the Michigan State Agricultural College, says: "I am delighted with it. I do not know of a more complete work on the subject, and I am pleased to state that it is filled with examples of the best methods of modern practice. We shall introduce it as a text-book in the college course." This is a fair specimen of the general reception.

Mathematical Almanac and Annual says:—

"Davies is a deservedly popular author, and his mathematical works are text-books in many of the leading schools and colleges."

Van Nostrand's Eclectic Engineering Magazine says:—

"We find in this new work all that can be asked for in a text-book. If there is a better work than this on Surveying, either for students or surveyors, our attention has not been called to it."

THE NEW LEGENDRE.

Van Amringe's Davies' Legendre.

Elements of Geometry and Trigonometry. By Charles Davies, LL.D. Revised (1885) by Prof. J. H. Van Amringe of Columbia College. New pages. 8vo. Full leather.

The present edition of the Legendre is the result of a careful re-examination of the work, into which have been incorporated such emendations in the way of greater clearness of expression or of proof as could be made without altering it in form or substance. Practical exercises are placed at the end of the several books, and comprise additional theorems, problems, and numerical exercises upon the principles of the Book or Books preceding. They will be found of great service in accustoming students, early in and throughout their course, to make for themselves practical application of geometric principles, and constitute, in addition, a large and excellent body of review and test questions for the convenience of teachers. The Trigonometry and mensuration have been carefully revised throughout; the deduction of principles and rules has been simplified; the discussion of the several cases which arise in the solution of triangles, plane and spherical, has been made more full and clear; and the whole has, in definition, demonstration, illustration, &c., been made to conform to the latest and best methods.

It is believed that in clearness and precision of definition, in general simplicity and rigor of demonstration, in the judicious arrangement of practical exercises, in orderly and logical development of the subject, and in compactness of form, Davies' Legendre is superior to any work of its grade for the general training of the logical powers of pupils, and for their instruction in the great body of elementary geometric truth.

The work has been printed from entirely new plates, and no care has been spared to **make it a model of typographical excellence.**

THE NATIONAL SERIES OF STANDARD SCHOOL-BOOKS.

DAVIES'S NATIONAL COURSE OF MATHEMATICS.

ITS RECORD.

In claiming for this series the first place among American text-books, of whatever class, the publishers appeal to the magnificent record which its volumes have earned during the *thirty-five years* of Dr. Charles Davies's mathematical labors. The unremitting exertions of a life-time have placed *the modern series* on the same proud eminence among competitors that each of its predecessors had successively enjoyed in a course of constantly improved editions, now rounded to their perfect fruition,—for it seems almost that this science is susceptible of no further demonstration.

During the period alluded to, many authors and editors in this department have started into public notice, and, by borrowing ideas and processes original with Dr. Davies, have enjoyed a brief popularity, but are now almost unknown. Many of the series of to-day, built upon a similar basis, and described as "modern books," are destined to a similar fate; while the most far-seeing eye will find it difficult to fix the time, on the basis of any data afforded by their past history, when these books will cease to increase and prosper, and fix a still firmer hold on the affection of every educated American.

One cause of this unparalleled popularity is found in the fact that the enterprise of the author did not cease with the original completion of his books. Always a practical teacher, he has incorporated in his text-books from time to time the advantages of every improvement in methods of teaching, and every advance in science. During all the years in which he has been laboring he constantly submitted his own theories and those of others to the practical test of the class-room, approving, rejecting, or modifying them as the experience thus obtained might suggest. In this way he has been able to produce an almost perfect series of class-books; in which every department of mathematics has received minute and exhaustive attention.

Upon the death of Dr. Davies, which took place in 1876, his work was immediately taken up by his former pupil and mathematical associate of many years, Prof. W. G. Peck, LL.D., of Columbia College. By him, with Prof. J. H. Van Amringe, of Columbia College, the original series is kept carefully revised and up to the times.

DAVIES'S SYSTEM IS THE ACKNOWLEDGED NATIONAL STANDARD FOR THE UNITED STATES, for the following reasons:—

1st. It is the basis of instruction in the great national schools at West Point and Annapolis.

2d. It has received the *quasi* indorsement of the National Congress.

3d. It is exclusively used in the public schools of the National Capital.

4th. The officials of the Government use it as authority in all cases involving mathematical questions.

5th. Our great soldiers and sailors commanding the national armies and navies were educated in this system. So have been a majority of eminent scientists in this country. All these refer to "Davies" as authority.

6th. A larger number of American citizens have received their education from this than from any other series.

7th. The series has a larger circulation throughout the whole country than any other, being *extensively used in every State in the Union*.

THE NATIONAL SERIES OF STANDARD SCHOOL-BOOKS.

DAVIES AND PECK'S ARITHMETICS.

OPTIONAL OR CONSECUTIVE.

The best thoughts of these two illustrious mathematicians are combined in the following beautiful works, which are the natural successors of Davies's Arithmetics, sumptuously printed, and bound in crimson, green, and gold:—

Davies and Peck's Brief Arithmetic.

Also called the "Elementary Arithmetic." It is the shortest presentation of the subject, and is *adequate* for all grades in common schools, being a thorough introduction to practical life, except for the specialist.

At first the authors play with the little learner for a few lessons, by object-teaching and kindred allurements; but he soon begins to realize that study is earnest, as he becomes familiar with the simpler operations, and is delighted to find himself master of important results.

The second part reviews the Fundamental Operations on a scale proportioned to the enlarged intelligence of the learner. It establishes the General Principles and Properties of Numbers, and then proceeds to Fractions. Currency and the Metric System are fully treated in connection with Decimals. Compound Numbers and Reduction follow, and finally Percentage with all its varied applications.

An Index of words and principles concludes the book, for which every scholar and most teachers will be grateful. How much time has been spent in searching for a half-forgotten definition or principle in a former lesson!

Davies and Peck's Complete Arithmetic.

This work certainly deserves its name in the best sense. Though complete, it is not, like most others which bear the same title, *cumbersome*. These authors excel in clear, lucid demonstrations, teaching the science pure and simple, yet not ignoring convenient methods and practical applications.

For turning out a thorough business man no other work is so well adapted. He will have a clear comprehension of the science as a whole, and a working acquaintance with details which must serve him well in all emergencies. Distinguishing features of the book are the logical progression of the subjects and the great variety of practical problems, not *puzzles*, which are beneath the dignity of educational science. A clear-minded critic has said o. Dr. Peck's work that it is free from that juggling with numbers which some authors falsely call "Analysis." A series of Tables for converting ordinary weights and measures into the Metric System appear in the later editions.

PECK'S ARITHMETICS.

Peck's First Lessons in Numbers.

This book begins with pictorial illustrations, and unfolds gradually the science of numbers. It noticeably simplifies the subject by developing the principles of addition and subtraction simultaneously; as it does, also, those of multiplication and division.

Peck's Manual of Arithmetic.

This book is designed especially for those who seek sufficient instruction to carry them successfully through practical life, but have not time for extended study.

Peck's Complete Arithmetic.

This completes the series but is a much briefer book than most of the complete arithmetics, and is recommended not only for what it contains, but also for what is omitted.

It may be said of Dr. Peck's books more truly than of any other series published, that they are clear and simple in definition and rule, and that superfluous matter of every kind has been faithfully eliminated, thus magnifying the working value of the book and saving unnecessary expense of time and labor.

BARNES'S NEW MATHEMATICS.

In this series JOSEPH FICKLIN, Ph. D., Professor of Mathematics and Astronomy in the University of Missouri, has combined all the best and latest results of practical and experimental teaching of arithmetic with the assistance of many distinguished mathematical authors.

Barnes's Elementary Arithmetic.
Barnes's National Arithmetic.

These two works constitute a *complete arithmetical course in two books*.

They meet the demand for text-books that will help students to acquire the greatest amount of useful and practical knowledge of Arithmetic by the smallest expenditure of *time, labor,* and *money*. Nearly every topic in Written Arithmetic is introduced, and its principles illustrated, by exercises in *Oral* Arithmetic. The free use of Equations; the concise method of combining and treating Properties of Numbers; the treatment of Multiplication and Division of Fractions in *two* cases, and then reduced to *one*; Cancellation by the use of the vertical line, especially in Fractions, Interest, and Proportion; the brief, simple, and greatly superior method of working Partial Payments by the "Time Table" and Cancellation; the substitution of formulas to a great extent for rules; the full and practical treatment of the Metric System, &c., indicate their completeness. A *variety* of methods and processes for the *same topic*, which deprive the pupil of the great benefit of doing a part of the *thinking* and *labor* for himself, have been discarded. The statement of principles, definitions, rules, &c., is brief and simple. The illustrations and methods are explicit, direct, and practical. The great number and variety of Examples embody the actual business of the day. The very large amount of matter condensed in so small a compass has been accomplished by economizing every line of space, by rejecting superfluous matter and obsolete terms, and by avoiding the *repetition* of analyses, explanations, and operations in the advanced topics which have been used in the more elementary parts of these books.

AUXILIARIES.

For use in district schools, and for supplying a text-book in advanced work for classes having finished the course as given in the ordinary Practical Arithmetics, the National Arithmetic has been divided and bound separately, as follows:—

Barnes's Practical Arithmetic.
Barnes's Advanced Arithmetic.

In many schools there are classes that for various reasons never reach beyond Percentage. It is just such cases where *Barnes's Practical Arithmetic* will answer a good purpose, at a *price to the pupil* much less than to buy the complete book. On the other hand, classes having finished the ordinary Practical Arithmetic can proceed with the higher course by using *Barnes's Advanced Arithmetic*.

For primary schools requiring simply a table book, and the earliest rudiments forcibly presented through object-teaching and copious illustrations, we have prepared

Barnes's First Lessons in Arithmetic,

which begins with the most elementary notions of numbers, and proceeds, by simple steps, to develop all the fundamental principles of Arithmetic.

Barnes's Elements of Algebra.

This work, as its title indicates, is elementary in its character and suitable for use, (1) in such public schools as give instruction in the Elements of Algebra; (2) in institutions of learning whose courses of study do not include Higher Algebra; (3) in schools whose object is to prepare students for entrance into our colleges and universities. This book will also meet the wants of students of Physics who require some knowledge of

Algebra. The student's progress in Algebra depends very largely upon the proper treatment of the four *Fundamental Operations*. The terms *Addition, Subtraction, Multiplication,* and *Division* in Algebra have a wider meaning than in Arithmetic, and these operations have been so defined as to *include* their arithmetical meaning; so that the beginner is simply called upon to *enlarge* his views of those fundamental operations. Much attention has been given to the explanation of the negative sign, in order to remove the well-known difficulties in the use and interpretation of that sign. Special attention is here called to "A Short Method of Removing Symbols of Aggregation," Art. 76. On account of their importance, the subjects of *Factoring, Greatest Common Divisor,* and *Least Common Multiple* have been treated at greater length than is usual in elementary works. In the treatment of *Fractions*, a method is used which is quite simple, and at the same time, more general than that usually employed. In connection with *Radical Quantities* the roots are expressed by fractional exponents, for the principles and rules applicable to integral exponents may then be used without modification. The *Equation* is made the chief subject of thought in this work. It is defined near the beginning, and used extensively in every chapter. In addition to this, four chapters are devoted exclusively to the subject of *Equations*. All *Proportions* are equations, and in their treatment as such all the difficulty commonly connected with the subject of Proportion disappears. The chapter on Logarithms will doubtless be acceptable to many teachers who do not require the student to master Higher Algebra before entering upon the study of Trigonometry.

HIGHER MATHEMATICS.

Peck's Manual of Algebra.
Bringing the methods of Bourdon within the range of the Academic Course.

Peck's Manual of Geometry.
By a method purely practical, and unembarrassed by the details which rather confuse than simplify science.

Peck's Practical Calculus.
Peck's Analytical Geometry.
Peck's Elementary Mechanics.
Peck's Mechanics, with Calculus.
The briefest treatises on these subjects now published. Adopted by the great Universities: Yale, Harvard, Columbia, Princeton, Cornell, &c.

Macnie's Algebraical Equations.
Serving as a complement to the more advanced treatises on Algebra, giving special attention to the analysis and solution of equations with numerical coefficients.

Church's Elements of Calculus.
Church's Analytical Geometry.
Church's Descriptive Geometry. With plates. 2 vols.
These volumes constitute the "West Point Course" in their several departments. Prof. Church was long the eminent professor of mathematics at West Point Military Academy, and his works are standard in all the leading colleges.

Courtenay's Elements of Calculus.
A standard work of the very highest grade, presenting the most elaborate attainable survey of the subject.

Hackley's Trigonometry.
With applications to Navigation and Surveying, Nautical and Practical Geometry, and Geodesy.

THE NATIONAL SERIES OF STANDARD SCHOOL-BOOKS.

BARNES'S ONE-TERM HISTORY SERIES.

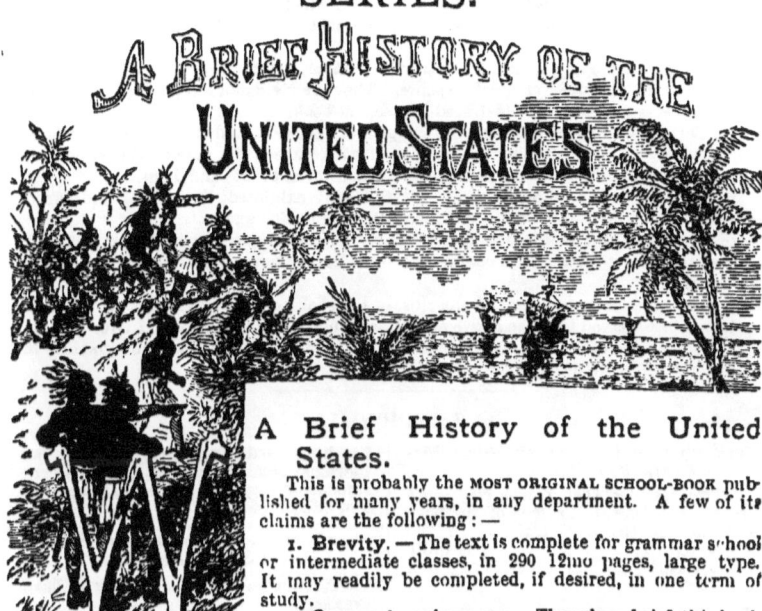

A Brief History of the United States.

This is probably the MOST ORIGINAL SCHOOL-BOOK published for many years, in any department. A few of its claims are the following: —

1. **Brevity.** — The text is complete for grammar school or intermediate classes, in 290 12mo pages, large type. It may readily be completed, if desired, in one term of study.

2. **Comprehensiveness.** — Though so brief, this book contains the pith of all the wearying contents of the larger manuals, and a great deal more than the memory usually retains from the latter.

3. **Interest** has been a prime consideration. Small books have heretofore been bare, full of dry statistics, unattractive. This one is charmingly written, replete with anecdote, and brilliant with illustration.

4. **Proportion of Events.** — It is remarkable for the discrimination with which the different portions of our history are presented according to their importance. Thus the older works, being already large books when the Civil War took place, give it less space than that accorded to the Revolution.

5. **Arrangement.** — In six epochs, entitled respectively, Discovery and Settlement, the Colonies, the Revolution, Growth of States, the Civil War, and Current Events.

6. **Catch Words.** — Each paragraph is preceded by its leading thought in prominent type, standing in the student's mind for the whole paragraph.

7. **Key Notes.** — Analogous with this is the idea of grouping battles, &c., about some central event, which relieves the sameness so common in such descriptions, and renders each distinct by some striking peculiarity of its own.

8. **Foot-Notes.** — These are crowded with interesting matter that is not strictly a part of history proper. They may be learned or not, at pleasure. They are certain in any event to be read.

9. **Biographies** of all the leading characters are given in full in foot-notes.

10. **Maps.** — Elegant and distinct maps from engravings on copper-plate, and beautifully colored, precede each epoch, and contain all the places named.

11. **Questions** are at the back of the book, to compel a more independent use of the text. Both text and questions are so worded that the pupil must give intelligent answers IN HIS OWN WORDS. "Yes" and "No" will not do.

THE NATIONAL SERIES OF STANDARD SCHOOL-BOOKS.

HISTORY — Continued.

12. Historical Recreations. — These are additional questions to test the student's knowledge, in review, as: "What trees are celebrated in our history?" "When did a fog save our army?" "What Presidents died in office?" "When was the Mississippi our western boundary?" "Who said, 'I would rather be right than President'?" &c.

13. The Illustrations, about seventy in number, are the work of our best artists and engravers, produced at great expense. They are vivid and interesting, and mostly upon subjects never before illustrated in a school-book.

14. Dates. — Only the leading dates are given in the text, and these are so associated as to assist the memory, but at the head of each page is the date of the event first mentioned, and at the close of each epoch a summary of events and dates.

15. The Philosophy of History is studiously exhibited, the causes and effects of events being distinctly traced and their inter-connection shown.

16. Impartiality. — All sectional, partisan, or denominational views are avoided. Facts are stated after a careful comparison of all authorities without the least prejudice or favor.

17. Index. — A verbal index at the close of the book perfects it as a work of reference.

It will be observed that the above are all particulars in which School Histories have been signally defective, or altogether wanting. Many other claims to favor it shares in common with its predecessors.

TESTIMONIALS.

From PROF. WM. F. ALLEN, *State University of Wisconsin.*

"Two features that I like *very much* are the *anecdotes* at the foot of the page and the '*Historical Recreations*' in the Appendix. The latter, I think, is quite a *new* feature, and the other is *very* well executed."

From HON. NEWTON BATEMAN, *Superintendent Public Instruction, Illinois.*

"Barnes's One-Term History of the United States is an exceedingly attractive and spirited little book. Its claim to several new and valuable features seems well founded. Under the form of six well-defined epochs, the history of the United States is traced tersely, yet pithily, from the earliest times to the present day. A good map precedes each epoch, whereby the history and geography of the period may be studied together, *as they always should be*. The syllabus of each paragraph is made to stand in such bold relief, by the use of large, heavy type, as to be of much *mnemonic* value to the student. The book is written in a sprightly and piquant style, the interest never flagging from beginning to end, — a rare and difficult achievement in works of this kind."

From HON. ABNER J. PHIPPS, *Superintendent Schools, Lewiston, Maine.*

"Barnes's History of the United States has been used for several years in the Lewiston schools, and has proved a very satisfactory work. I have examined the new edition of it."

From HON. R. K. BUCHELL, *City Superintendent Schools, Lancaster, Pa.*

"It is the *best* history of the kind I have ever seen."

From T. J. CHARLTON, *Superintendent Public Schools, Vincennes, Ind.*

"We have used it here for six years, and it has given almost perfect satisfaction. ... The notes in fine print at the bottom of the pages are of especial value."

From PROF. WM. A. MOWRY, *E. & C. School, Providence, R. I.*

"Permit me to express my high appreciation of your book. I wish all text-books for the young had equal merit."

From HON. A. M. KEILEY, *City Attorney, Late Mayor, and President of the School Board, City of Richmond, Va.*

"I do not hesitate to volunteer to you the opinion that Barnes's History is entitled to the preference in almost every respect that distinguishes a good school-book. ... The narrative generally exhibits the temper of the judge; rarely, if ever, of the advocate."

THE NATIONAL SERIES OF STANDARD SCHOOL-BOOKS.

Primary History of the United States.

For Intermediate Classes. 12mo. 225 pages. Beautifully illustrated. A fitting introduction to Barnes's Historical Series.

From PROF. C. W. RICHARDS, *High School, Oswego, N.Y.*

"I think it an admirable book."

From D. BEACH, *of Gibbons & Beach, 20 West 59th Street, N.Y. City.*

"The little History is to me a very attractive book."

From PROF. C. D. LARKINS, *Fayetteville, N.Y.*

"It is the only Primary History that I ever saw that I liked."

From PROF. L. R. HOPKINS, *Weedsport, N.Y.*

"I think Barnes's Primary History by far the best I ever saw."

From PROF. RICHARD H. LEWIS, *Kingston College, N.C.*

"The subject matter is very good, and shows remarkable condensing power in the author."

From PROF. EDWARD SMITH, *Supt. of Schools, Syracuse, N.Y.*

"It is a very interesting and pretty book. I should like it very much for supplementary reading."

From GENERAL HORATIO C. KING, *Brooklyn, N.Y.*

"I am especially pleased with the new Primary History, which is remarkably concise and interesting and free from partisan bias."

From PROF S. G. HARRIS, *Dryden, N.Y.*

"Having a few days' vacation I found time to carefully examine the Primary History you sent me and am highly delighted with it. It will satisfy a long-felt want."

From the New England Journal of Education.

"The book is printed in the best type, on the finest paper, and is illustrated in the most superb, even sumptuous manner. Any child who studies this exceptionally beautiful little book will unavoidably have a higher regard for his country on account of the superior and charming character of the book."

From MR. H. H. SMITH, *Prest. Board of Education, Vineyard Haven, Mass.*

"I should think you would feel proud of the work."

From DR. EUGENE BOUTON, *Albany, N.Y.*

"I must congratulate every one on the publication of this beautiful History."

From PROF. H. C. TALMADGE, *Woodbury, Ct.*

"It is the book that I have been looking for quite a long time."

From PROF. L. C. FOSTER, *Supt. of Schools, Ithaca, N.Y.*

"It is indeed a very beautiful book, and it seems to me well adapted for use in the lower grammar grades."

From PROF. F. H. HALL, *Sinclairville, N.Y.*

"This History is the best thing of the kind I have ever seen. How it could be improved I do not see."

From PROF. J. C CRUIKSHANK, *Supt. of Education, Passaic Co., N.J.*

"It is the book needed, and will fill the gap of early historical instruction in the schools."

From PROF. S. R. MORSE, *Supt. of Education, Atlantic Co., N.J.*

"I have examined Barnes's Primary History of the United States and find it just what we have wanted in our schools."

From H. E. PERKINS, *School Commissioner, Livingston Co., N.Y.*

"I think it the best Primary United States History that I ever examined, and will recommend it to my teachers."

From The Indiana School Journal.

"This book, comprised in 225 pages, is what its title indicates, primary in matter and manner of treatment, and not simply an abbreviation of a large book. By not attempting everything there is space for a fuller discussion of the more important points. The author has clearly discriminated between simplicity of style and simple thought."

THE NATIONAL SERIES OF STANDARD SCHOOL-BOOKS.

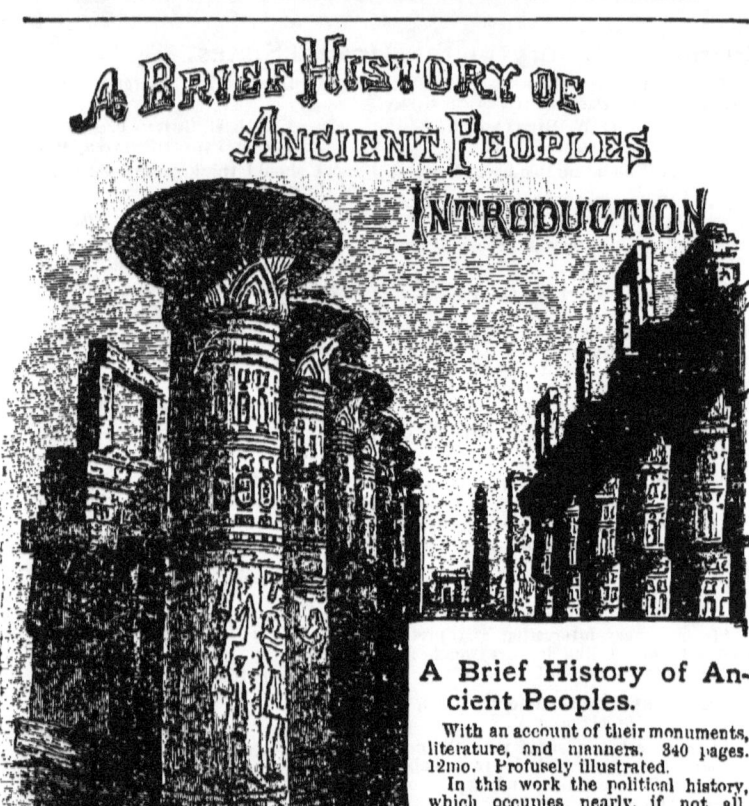

A Brief History of Ancient Peoples Introduction

A Brief History of Ancient Peoples.

With an account of their monuments, literature, and manners. 340 pages. 12mo. Profusely illustrated.

In this work the political history, which occupies nearly, if not all, the ordinary school text, is condensed to the salient and essential facts, in order to give room for a clear outline of the literature, religion, architecture, character, habits, &c., of each nation. Surely it is as important to know *something* about Plato as *all* about Cæsar, and to learn how the ancients wrote their books as how they fought their battles.

The chapters on Manners and Customs and the Scenes in Real Life represent the people of history as men and women subject to the same wants, hopes and fears as ourselves, and so bring the distant past near to us. The Scenes, which are intended *only for reading*, are the result of a careful study of the unequalled collections of monuments in the London and Berlin Museums, of the ruins in Rome and Pompeii, and of the latest authorities on the domestic life of ancient peoples. Though intentionally written in a semi-romantic style, they are accurate pictures of what *might* have occurred, and some of them are simple transcriptions of the details sculptured in Assyrian alabaster or painted on Egyptian walls.

HISTORY — *Continued.*

The extracts made from the sacred books of the East are not specimens of their style and teachings, but only gems selected often from a mass of matter, much of which would be absurd, meaningless, and even revolting. It has not seemed best to cumber a book like this with selections conveying no moral lesson.

The numerous cross-references, the abundant dates in parenthesis, the pronunciation of the names in the Index, the choice reading references at the close of each general subject, and the novel Historical Recreations in the Appendix, will be of service to teacher and pupil alike.

Though designed primarily for a text-book, a large class of persons — general readers, who desire to know something about the progress of historic criticism and the recent discoveries made among the resurrected monuments of the East, but have no leisure to read the ponderous volumes of Brugsch, Layard, Grote, Mommsen, and Ihne — will find this volume just what they need.

From HOMER B. SPRAGUE, *Head Master Girls' High School, West Newton St., Boston, Mass.*

"I beg to recommend in strong terms the adoption of Barnes's 'History of Ancient Peoples' as a text-book. It is about as nearly perfect as could be hoped for. The adoption would give great relish to the study of Ancient History."

HE Brief History of France.

By the author of the "Brief United States," with all the attractive features of that popular work (which see) and new ones of its own.

It is believed that the History of France has never before been presented in such brief compass, and this is effected without sacrificing one particle of interest. The book reads like a romance, and, while drawing the student by an irresistible fascination to his task, impresses the great outlines indelibly upon the memory.

30

THE NATIONAL SERIES OF STANDARD SCHOOL-BOOKS.

HISTORY — *Continued.*

Barnes's Brief History of Mediæval and Modern Peoples.

The success of the History of Ancient Peoples was immediate and great. A History of Mediæval and Modern History, upon the same plan, was the natural sequence. Those teachers who used the former will be glad to know that the latter book is now ready, and classes can go right on without changing authors.

The New York *School Journal* says : —
"The fine-print notes ... work a field not widely developed until Green's History of English People appeared, relating to the description of real, every-day life of the people."

This work distinguishes between the period of the world's history from the Fall of Rome (A.D. 476) to the Capture of Constantinople (A.D. 1453),—about one thousand years, called "Middle Ages,"— and the period from the end of the fifteenth century to the present time. It covers the entire time chronologically and by the order of events, giving one hundred and twenty-two fine illustrations and sixteen elaborate maps.

[Illustration from Barnes's Brief-History Series.]

The subject has never before been so interestingly treated in brief compass. The Political History of each nation is first given, then the Manners and Customs of the People. A better idea of the growth of civilization and the changes in the condition of mankind cannot be found elsewhere. The book is fitted for private reading, as well as schools.

THE NATIONAL SERIES OF STANDARD SCHOOL-BOOKS.

HISTORY — Continued.

Barnes's Brief General History.

Comprising Ancient, Mediæval, and Modern Peoples.

THE SPECIAL FEATURES OF THIS BOOK ARE AS FOLLOWS: —

The **General History** contains 600 pages. Of this amount, 350 pages are devoted to the political history, and 250 pages to the civilization, manners, and customs, etc. The latter are in separate chapters, and if the time of the teacher is limited, may be omitted. The class can thus take only the political portion when desired. The teacher will have, however, the satisfaction of knowing that, such is the fascinating treatment of the civilization, literature, etc., those chapters will be carefully read by the pupils; and, on the principle that knowledge acquired from love alone is the most vivid, will probably be the best-remembered part of the book. This portion of the book is therefore all clear gain.

The **Black-board Analysis.** See p. 314 as an example of this marked feature.

The exquisite **Illustrations**, unrivalled by any text-book. See pp. 9, 457, and 582, as samples of the 240 cuts contained in this beautiful work.

The peculiar **Summaries**, and valuable lists of **Reading References**. See p. 417.

The numerous and excellent colored **Maps.** These are so full as to answer for an extensive course of collateral reading, and are consequently useful for reference outside of class-work. See pp. 299 and 317.

The **Scenes in Real Life**, which are the result of a careful study of the collections and monuments in the London, Paris, and Berlin museums, and the latest authorities upon the domestic life of the people of former times. See pp. 38-39. This scene — a Lord of the IVth Dynasty — is mainly a transcription of details to be found painted on the walls of Egyptian tombs.

The chapters on **Civilization** that attempt to give some idea of the Monuments, Arts, Literature, Education, and Manners and Customs of the different nations. See pp. 171, 180, 276, 279, 472, and 514.

The admirable **Genealogical Tables** interspersed throughout the text. See pp. 340 and 404.

The **Foot-Notes** that are packed full of anecdotes, biographies, pleasant information, and suggestive comments. As an illustration of these, take the description of the famous sieges of Haarlem and Leyden, during the Dutch War of Independence, pp. 446 and 448.

The peculiar method of treating **Early Roman History**, by putting in the text the facts as accepted by critics, and, in the notes below, the legends. See pp. 205-6.

The exceedingly useful plan of running collateral history in parallel columns, as for example on p. 361, taken from the Hundred Years' War.

The **Historical Recreations**, so valuable in arousing the interest of a class. See p. xi from the Appendix.

The striking opening of Modern History on pp. 423-4.

The interesting **Style**, that sweeps the reader along as by the fascination of a novel. The pupil insensibly acquires a taste for historical reading, and forgets the tediousness of the ordinary lesson in perusing the thrilling story of the past. See pp. 251-2.

Special attention is called to the chapter entitled **Rise of Modern Nations,** — England, France, and Germany. The characteristic feature in the mediæval history of each of these nations is made prominent. (a.) After the Four Conquests of England, the central idea in the growth of that people was the Development of Constitutional Liberty. (b.) The feature of French history was the conquest of the great vassals by the king, the triumph of royalty over feudalism, and the final consolidation of the scattered fiefs into one grand monarchy. (c.) The characteristic of German history was disunion, emphasized by the lack of a central capital city, and by an elective rather than an hereditary monarchy. The struggle of the Crown with its powerful vassals was the same as in France, but developed no national sentiment, and ended in the establishment of semi-independent dukedoms.

These three thoughts furnish the beginner with as many threads on which to string the otherwise isolated facts of this bewildering period.

THE NATIONAL SERIES OF STANDARD SCHOOL-BOOKS.

BOOK-KEEPING.

Powers's Practical Book-keeping.
Powers's Blanks to Practical Book-keeping.

A Treatise on Book-keeping, for Public Schools and Academies. By Millard R. Powers, M. A. This work is designed to impart instruction upon the science of accounts, as applied to mercantile business, and it is believed that more knowledge, and that, too, of a more practical nature, can be gained by the plan introduced in this work, than by any other published.

Folsom's Logical Book-keeping.
Folsom's Blanks to Book-keeping.

This treatise embraces the interesting and important discoveries of Professor Folsom (of the Albany " Bryant & Stratton College "), the partial enunciation of which in lectures and otherwise has attracted so much attention in circles interested in commercial education.

After studying business phenomena for many years, he has arrived at the positive laws and principles that underlie the whole subject of accounts ; finds that the science is based in *value* as a generic term ; that value divides into *two classes* with varied species ; that all the exchanges of values are reducible to nine equations ; and that all the results of all these exchanges are limited to *thirteen* in number.

As accounts have been universally taught hitherto, without setting out from a radical analysis or definition of values, the science has been kept in great obscurity, and been made as difficult to impart as to acquire. On the new theory, however, these obstacles are chiefly removed. In reading over the first part of it, in which the governing laws and principles are discussed, a person with ordinary intelligence will obtain a fair conception of the *double-entry* process of accounts. But when he comes to study thoroughly these laws and principles as there enunciated, and works out the examples and memoranda which elucidate the *thirteen results* of business, the student will neither fail in readily acquiring the science as it is, nor in becoming able intelligently to apply it in the interpretation of business.

Smith and Martin's Book-keeping.
Smith and Martin's Blanks.

This work is by a practical teacher and a practical book-keeper. It is of a thoroughly popular class, and will be welcomed by every one who loves to see theory and practice combined in an easy, concise, and methodical form.

The single-entry portion is well adapted to supply a want felt in nearly all other treatises, which seem to be prepared mainly for the use of wholesale merchants ; leaving retailers, mechanics, farmers, &c., who transact the greater portion of the business of the country, without a guide. The work is also commended, on this account, for general use in young ladies' seminaries, where a thorough grounding in the simpler form of accounts will be invaluable to the future housekeepers of the nation.

The treatise on double-entry book-keeping combines all the advantages of the most recent methods with the utmost simplicity of application, thus affording the pupil all the advantages of actual experience in the counting-house, and giving a clear comprehension of the entire subject through a judicious course of mercantile transactions.

PRACTICAL BOOK-KEEPING.

Stone's Post-Office Account Book.

By Micah H. Stone. For record of Box Rents and Postages. Three sizes always in stock. 64, 108, and 204 pages.

INTEREST TABLES.

Brooks's Circular Interest Tables.

To calculate simple and compound interest for any amount, from 1 cent to $1,000, at current rates from 1 day to 7 years.

DR. STEELE'S ONE-TERM SERIES, IN ALL THE SCIENCES.

Steele's 14-Weeks Course in Chemistry.
Steele's 14-Weeks Course in Astronomy.
Steele's 14-Weeks Course in Physics.
Steele's 14-Weeks Course in Geology.
Steele's 14-Weeks Course in Physiology.
Steele's 14-Weeks Course in Zoölogy.
Steele's 14-Weeks Course in Botany.

Our text-books in these studies are, as a general thing, dull and uninteresting. They contain from 400 to 600 pages of dry facts and unconnected details. They abound in that which the student cannot learn, much less remember. The pupil commences the study, is confused by the fine print and coarse print, and neither knowing exactly what to learn nor what to hasten over, is crowded through the single term generally assigned to each branch, and frequently comes to the close without a definite and exact idea of a single scientific principle.

Steele's "Fourteen-Weeks Courses" contain only that which every well-informed person should know, while all that which concerns only the professional scientist is omitted. The language is clear, simple, and interesting, and the illustrations bring the subject within the range of home life and daily experience. They give such of the general principles and the prominent facts as a pupil can make familiar as household words within a single term. The type is large and open; there is no fine print to annoy; the cuts are copies of genuine experiments or natural phenomena, and are of fine execution.

In fine, by a system of condensation peculiarly his own, the author reduces each branch to the limits of a single term of study, while sacrificing nothing that is essential, and nothing that is usually retained from the study of the larger manuals in common use. Thus the student has rare opportunity to *economize his time*, or rather to employ that which he has to the best advantage.

A notable feature is the author's charming "style," fortified by an enthusiasm over his subject in which the student will not fail to partake. Believing that Natural Science is full of fascination, he has moulded it into a form that attracts the attention and kindles the enthusiasm of the pupil.

The recent editions contain the author's "Practical Questions" on a plan never before attempted in scientific text-books. These are questions as to the nature and cause of common phenomena, and are not directly answered in the text, the design being to test and promote an intelligent use of the student's knowledge of the foregoing principles.

Steele's Key to all His Works.

This work is mainly composed of answers to the Practical Questions, and solutions of the problems, in the author's celebrated "Fourteen-Weeks Courses" in the several sciences, with many hints to teachers, minor tables, &c. Should be on every teacher's desk.

Prof. J. Dorman Steele is an indefatigable student, as well as author, and his books have reached a fabulous circulation. It is safe to say of his books that they have accomplished more tangible and better results in the class-room than any other ever offered to American schools, and have been translated into more languages for foreign schools. They are even produced in raised type for the blind.

NATURAL SCIENCE — *Continued.*

TEMPERANCE PHYSIOLOGY.

Steele's Abridged Physiology, for Common Schools.
Steele's Hygienic Physiology, for High Schools.

With especial reference to alcoholic drinks and narcotics. Adapted from "Fourteen Weeks' Course in Human Physiology." By J. Dorman Steele, Ph.D. Edited and indorsed for the use of schools (in accordance with the recent legislation upon this subject) by the Department of Temperance Instruction of the W. C. T. U. of the United States, under the direction of Mrs. Mary H. Hunt, superintendent.

This new work contains all the excellent and popular features that have given Dr. Steele's Physiology so wide a circulation. Among these, are the following:

1. **Colored Lithographs** to illustrate the general facts in Physiology.
2. **Black-board Analysis** at the beginning of each chapter. These have been found of great service in class-work, especially in review and examination.
3. The **Practical Questions** at the close of each chapter. These are now too well known to require any explanation.
4. The carefully prepared sections upon the **Physiological Action of Alcohol, Tobacco, Opium,** etc. These are scattered through the book as each organ is treated. This subject is examined from a purely scientific stand-point, and represents the latest teachings at home and abroad. While there is no attempt to incorporate a temperance lecture in a school-book, yet the terrible effects of these "Stimulants and Narcotics," especially upon the young, are set forth all the more impressively, since the lesson is taught merely by the presentation of facts that lean toward no one's prejudices, and admit of no answer or escape.
5. Throughout the book, there are given, in text and foot-note, experiments that can be performed by teacher and pupil, and which, it is hoped, will induce some easy dissections to be made in every class, and lead to that constant reference of all subjects to Nature herself, which is so invaluable in scientific study.
6. The collection of **recent discoveries, interesting facts,** etc., in numerous foot-notes.
7. The unusual space given to the subject of **Ventilation,** which is now attracting so much attention throughout the country.
8. The text is brought up to the level of the new Physiological views. The division into short, pithy paragraphs; the bold paragraph headings; the clear, large type; the simple presentation of each subject; the interesting style that begets in every child a love of the study, and the beautiful cuts, each having a full scientific description and nomenclature, so as to present the thing before the pupil without cumbering the text with the dry details, — all these indicate the work of the practical teacher, and will be appreciated in every school-room.

Child's Health Primer.
For the youngest scholars. 12mo, cloth, illustrated.

Hygiene for Young People.
Prepared under the supervision of Mrs. Mary H. Hunt, Superintendent of the Department of Scientific Instruction of the "Women's National Christian Temperance Union." Examined and approved by A. B. Palmer, M.D., University of Michigan.

Jarvis's Elements of Physiology.
Jarvis's Physiology and Laws of Health.

The only books extant which approach this subject with a proper view of the true object of teaching Physiology in schools, viz., that scholars may know how to take care of their own health. In bold contrast with the abstract *Anatomies,* which children learn as they would Greek or Latin (and forget as soon), to *discipline the mind,* are these text-books, using the *science* as a secondary consideration, and only so far as is necessary for the comprehension of the *laws of health.*

www.ingramcontent.com/pod-product-compliance
Lightning Source LLC
Chambersburg PA
CBHW020239240426
43672CB00006B/573